Springer Briefs in Archaeology

SpringerBriefs in Underwater Archaeology

Series Editor
Annalies Corbin, Columbus, OH, USA

Books in the SpringerBriefs in Underwater Archaeology series, published in cooperation with the Advisory Council on Underwater Archaeology (ACUA), will address critical contemporary problems and illustrate exemplary work in maritime cultural heritage in countries around the globe. The series will take a broad view of the theory, concepts, issues and complexities associated with the management and protection of maritime cultural heritage sites in the 21st century. The book series will be concerned with: • submerged cultural resources ranging in age from the most ancient human history through the modern era • maritime cultural heritage, understood as not just archaeological sites but also the stakeholders who inevitably constitute diverse and multi-scalar sectors of society • management that transcends adherence simply to regulation but that looks forward to the needs of the future • emerging technologies that will transform the ways in which archaeological research is conducted and archaeological resources, both sites and landscapes, are understood and interpreted • site stewardship and dissemination of information that serves the needs of the archaeological field and the public at large. All volumes will be stringently peer-reviewed, first by the ACUA series editor and associate editors, and upon passing that level, two outside reviewers familiar and qualified to make constructive comments regarding the manuscript.

Michael L. Brennan
Editor

Threats to Our Ocean Heritage: Potentially Polluting Wrecks

2021 United Nations Decade
2030 of Ocean Science for Sustainable Development

THE OCEAN FOUNDATION

Springer

Editor
Michael L. Brennan
SEARCH Inc.
Jacksonville, FL, USA

ISSN 1861-6623   ISSN 2192-4910   (electronic)
SpringerBriefs in Archaeology
ISSN 2625-2562   ISSN 2625-2570 (electronic)
SpringerBriefs in Underwater Archaeology
ISBN 978-3-031-57959-2   ISBN 978-3-031-57960-8   (eBook)
https://doi.org/10.1007/978-3-031-57960-8

This work was supported by Ocean Foundation

© The Editor(s) (if applicable) and The Author(s) 2024. This book is an open access publication.
Open Access This book is licensed under the terms of the Creative Commons Attribution 4.0
International License (http://creativecommons.org/licenses/by/4.0/), which permits use, sharing,
adaptation, distribution and reproduction in any medium or format, as long as you give appropriate credit
to the original author(s) and the source, provide a link to the Creative Commons license and indicate if
changes were made.
The images or other third party material in this book are included in the book's Creative Commons
license, unless indicated otherwise in a credit line to the material. If material is not included in the book's
Creative Commons license and your intended use is not permitted by statutory regulation or exceeds the
permitted use, you will need to obtain permission directly from the copyright holder.
The use of general descriptive names, registered names, trademarks, service marks, etc. in this publication
does not imply, even in the absence of a specific statement, that such names are exempt from the relevant
protective laws and regulations and therefore free for general use.
The publisher, the authors and the editors are safe to assume that the advice and information in this book
are believed to be true and accurate at the date of publication. Neither the publisher nor the authors or the
editors give a warranty, expressed or implied, with respect to the material contained herein or for any
errors or omissions that may have been made. The publisher remains neutral with regard to jurisdictional
claims in published maps and institutional affiliations.

This Springer imprint is published by the registered company Springer Nature Switzerland AG
The registered company address is: Gewerbestrasse 11, 6330 Cham, Switzerland

If disposing of this product, please recycle the paper.

Foreword

The Ocean Foundation prioritises protecting cultural heritage as part of its ocean conservation goals. Globally, we share a common ocean heritage (both natural and cultural). The Ocean Foundation is part of a UN Decade of Ocean Science for a Sustainable Development endorsed project. This Ocean Decade project focuses on the sustainable development of our 'Ocean Heritage' so that fishing, mining, and the coastal environment are safer for individuals, businesses, and societies, and more sustainable for those dependent upon a healthy ocean for their life and livelihoods. These social and cultural impacts are closely aligned with the public education focus of the mission of our partner, Lloyd's Register Foundation. The project is also in line with the safety challenge areas, such as the Safety of Food or the provision of Safe and Sustainable Systems from fishing, mining, and salvage of potentially polluting wrecks. The impact will be a greater degree of ocean literacy in several topics across a wide range of stakeholders.

The 1972 UN Convention concerning the Protection of the World Cultural and Natural Heritage emerged from two separate movements: one focusing on the preservation of cultural sites, and the other dealing with the conservation of nature. 'Natural Heritage' in this case is largely synonymous with natural resources and includes biodiversity and the ecosystem, as well as geological structures around which biodiversity thrives.

In alignment with the 1972 Convention, we at The Ocean Foundation seek to protect both cultural and natural World Heritage sites. From the dugongs of Okinawa, to the wreck of *Titanic*, and the grey whale nursery and coastal heritage of Laguna San Ignacio, we have looked for ways to support communities, biodiversity, and history. At the same time, we know that it is important to not only document and honour wreck sites to the extent possible, but also to manage them for the benefit of living marine resources and the human communities that depend on them.

Military wrecks are deemed protected as maritime gravesites (and legally belong forever to the home nation). In some cases, it was easy to provide *in situ* protection of wrecks in places such as Chuuk Lagoon in the Federated States of Micronesia. There, new sites are being discovered, helping to solve the mystery of still-missing service personnel, and enhancing the Lagoon's role as a dive tourism destination. In addition, the wrecks are co-located with amazing natural underwater resources which makes protection of both cultural and natural heritage a community benefit.

In this book, the authors shed light on the *unintended* consequences of protecting shipwrecks *in situ* as cultural heritage—particularly when that protection may end up posing complex environmental, social, and economic hazards. This is because these wrecks sometimes sank still loaded with bunker fuel, oils, chemical warfare agents, and unexploded munitions. Therefore, historic resources that are often war graves also pose environmental hazards that must be mitigated while also taking into account the integrity of the submerged cultural resource.

Sunken wrecks are not stable. And, after eight decades or more on the seafloor, these wrecks are subject to metal fatigue, corrosion, and even implosion under the weight of the water above them. Increasingly intense and frequent storms accelerate the break-up process. This creates the opportunity for the petroleum cargoes and fuel still on board to leak and pollute the local environment. Shifting wreck debris itself can damage nearby coral structures, seagrass meadows, mangroves, and salt marshes. As such, the wrecks can become a threat to the associated fisheries, dive tourism, and the health of humans and marine life alike.

Any one leak may become a locally significant source of pollution. The IUCN estimates there are 8,500 wrecks out there that are at risk of leaking as much as 6 billion gallons of oil; the equivalent of 545 Exxon Valdez spills.

As this book will make clear, there is not enough data on the status of those 8,500 sunken ships, not to mention planes, tanks, trucks, barrels, or other cargo they may have been carrying. Too often, governments, communities, and conservationists do not discover a leak until it is causing harm. But we do know it is less expensive to prevent pollution than it is to restore damaged ecosystems.

Coastal communities, and the ecosystems they depend upon, would benefit from better international cooperation on mapping wrecks, evaluating risks, sharing pollution prevention methodologies, and prioritising limited resources to address the greatest risks first. And, given the challenges of higher ocean temperatures, increasing water depth, deoxygenation, and acidification, it behooves nations to work together to ensure that we embrace the natural cultural heritage that our ocean provides and prevent any further harm.

The Ocean Foundation   Mark J. Spalding
Washington, DC, USA

Contents

Abbreviations

ABNJ	Areas Beyond National Jurisdiction
AF	Air France
AIS	Automatic Identification System
bbl	barrel
BBNJ	Biodiversity Beyond National Jurisdiction
BDoP	Polish Database of Underwater Objects
BNSC	British National Space Centre
BOEM	Bureau of Ocean Energy Management
BP	British Petroleum
BSH	German Federal Maritime and Hydrographic Agency
BTEX	Benzene, Toluene, Ethylbenzene, and Xylene
CECR	Concretion Equivalent Corrosion Rate
CERCLA	Comprehensive Environmental Response, Compensation and Liability Act
cm	centimeter
CS	Continental Shelf
DEEPP	Development of European Guideline for Potentially Polluting Shipwrecks
Defra	Department for Environment, Food and Rural Affairs
DIP	Diplomatic
DOI	Department of the Interior
DMSA	Danish Maritime Safety Authority
E corr.	Electro-reactivity potential
E-DBA	Environmental Desk-Based Assessment
EEZ	Exclusive Economic Zone
EFC	Emergency Fleet Corporation
EIA	Environmental Impact Assessment
EPA	Environmental Protection Agency
EU	European Union
E/V	Exploration Vessel
EWT	Eastern War Time

ft	foot
FEA	Finite Element Analysis
FMIS	Informationssystemet över fornminnen
FOSC	Federal On-Scene Coordinator
FSM	Federated States of Micronesia
g	gram
GIOS	Polish Chief Inspectorate for Environmental Protection
GIS	Geographic Information System
GMT	Greenwich Mean Time
GOA	Government of Australia
GOMRI	Gulf of Mexico Research Initiative
GPS	Global Positioning System
GRMI	Government of the Republic of the Marshall Islands
GT	Gross Tons
HELCOM	Helsinki Commission
HMAS	His/Her Majesty's Australian Ship
HMS	His/Her Majesty's Ship
HNLMS	His/Her Netherland Majesty's Ship
ICOMOS	International Council on Monuments and Sites
IMG	Maritime Institute in Gdansk
IMO	International Maritime Organization
IMUMG	Maritime Institute of Maritime University of Gdynia
IOPAS	Institute of Oceanology Polish Academy of Science
ISA	International Seabed Authority
ITLOS	International Tribunal for the Law of the Sea
IUCN	International Union for the Conservation of Nature
JMAS	Japanese Mine Action Service
km	kilometer
KptLt	Kapitänleutnant
kt	knot
L	Liter
LC1	Lethal Concentration 1
LC50	Lethal Concentration 50
LOSC	Law of the Sea Convention
m	meter
MAH	Mono-Cyclic Aromatic Hydrocarbons
MBES	Multibeam echosounder
MIC	Microbiologically Influenced Corrosion
mm	millimeter
MOD	Ministry of Defence
MPA	Marine Protected Area
MPF	Major Projects Foundation
mph	miles per hour
mt	metric tons
MV	Motor Vessel

NBS	Neutron Backscatter System
NGO	Non-Governmental Organisation
NHPA	National Historic Preservation Act
nm	nautical mile
NOAA	National Oceanic and Atmospheric Administration
NRC	National Research Council
NRT	Navigation Response Team
NRDA	Natural Resource Damage Assessments
NRDAM/CME	Natural Resource Damage Assessment Model for Coastal and Marine Environments
OCNO	Chief of Naval Operations
OCP	Open Circuit Potential
OEEM	Office of Environment and Emergency Management
OET	Ocean Exploration Trust
ONMS	Office of National Marine Sanctuaries
PACPLAN	Pacific Islands Regional Marine Spill Contingency Plan
PACPOL	Pacific Ocean Pollution Prevention Programme
PAH	Poly-cyclic Aromatic Hydrocarbons
ΣPAH	summed PAH
PALM	Pacific Islands Leaders Meeting
PMRA	Protection of Military Remains Act
PPW	Potentially Polluting Wreck
RAMP	Rapid Ecological Assessment and Monitoring
RFA	Royal Fleet Auxiliary
RMI	Republic of the Marshall Islands
RMS	Royal Mail Ship
ROV	Remotely Operated Vehicle
RRT	Regional Response Team
RULET	Remediation of Underwater Legacy Environmental Threats
RUST	Resources and UnderSea Threats
SALMO	Salvage and Marine Operations
SAR	Search and Rescue
SAR	Synthetic Aperture Radar
S-boat	Schnellboot (fast boat)
SHIELDS	Sanctuaries Hazards Incident Emergency Logistics Database System
SHIPWHER	Estonian central wreck register
SIMAP	Spill Impact Model Application Package
SMCA	Sunken Military Craft Act
SOPAC	South Pacific Applied Geoscience Commission
SPREP	Secretariat of the Pacific Regional Environment Programme
SS	Steamship
SUPSALV	US Navy Supervisor of Salvage
SWERA	Solar and Wind Energy Assessment
SYKE	Finnish Environment Institute

THC	Total Hydrocarbon Content
UASBC	Underwater Archaeological Society of British Columbia
U-boat	Unterseeboot (under-sea boat)
UCH	Underwater Cultural Heritage
µg	micrograms
UK	United Kingdom
UN	United Nations
UNCLOS	UN Conference on the Law of the Sea
UNESCO	United Nations Educational, Scientific and Cultural Organization
USD	United States dollar
USCG	U.S. Coast Guard
USG	U.S. Government
USINDOPACOM	U.S. Indo-Pacific Command
USNS	United States Naval Ship
US	United States
USS	United States Ship
VTS	Vessel Traffic Services
WCD	Worst Case Discharge
WHC	World Heritage Convention
WWI	World War I
WWII	World War II

Chapter 1
Potentially Polluting Wrecks: An Introduction

Michael L. Brennan

1.1 Overview

Shipwrecks lie hidden below the surface of the water and, especially those in deep water, are out of sight and easily ignored. Many of these wrecks from the modern era either contain or are suspected to contain hazardous materials that are within the metal hulls which have the potential to cause an environmental disaster should they leak or spill. A potentially polluting wreck (PPW) is a shipwreck containing a cargo or a large volume of its own fuel that remains within the wreck and has the potential to cause an environmental hazard should the structure become compromised and either leak or catastrophically release. This book addresses those wrecks with the potential to pollute due to petroleum cargoes or bunkers. While unexploded ordnance and munitions also represent both a hazard and toxic substance, addressing this type of cargo and threat is different than for petroleum cargos and would need its own volume on the subject. According to the International Union for the Conservation of Nature (IUCN), 'marine pollution from sunken vessels is predicted to reach its highest level this decade, with over 8,500 shipwrecks at risk of leaking' (IUCN, 2023). They contain hazardous materials including chemicals, unexploded ordinances, and an estimated six billion gallons of heavy fuel oil. 'This is 545 times more oil than the Exxon Valdez leak in 1989 and 30 times that of the *Deepwater Horizon* spill in 2010, both of which had severe and long-lasting environmental consequences' (IUCN, 2023). Many of the wrecks identified as PPWs are those sunk during the two world wars, particularly oil tankers, but also include freighters, as well as ships from parts of the twentieth century that foundered in storms.

M. L. Brennan (✉)
SEARCH Inc., Jacksonville, FL, USA
e-mail: mike@brennanexploration.com

© The Author(s) 2024
M. L. Brennan (ed.), *Threats to Our Ocean Heritage: Potentially Polluting Wrecks*, SpringerBriefs in Underwater Archaeology, https://doi.org/10.1007/978-3-031-57960-8_1

A study in 2005 for the International Oil Spill Conference compiled a worldwide dataset of PPWs and identified 8569 potentially polluting wrecks with the parameters of tankers greater than 150 gross tons (GT) (1583) and non-tank vessels greater than 400 GT (6986) (Michel et al., 2005: 69). The potential environmental risks these wrecks pose are significant. The common example of SS *Jacob Luckenbach*, sunk in a collision off San Francisco's Golden Gate in 1953, was estimated to have killed thousands of birds over the course of a decade of mystery oil slicks from the wreck with untold other impacts to the marine environment. Only in the aftermath of the *Deepwater Horizon* spill and the research conducted in the Gulf of Mexico since do we have a better understanding of some of the environmental impacts of such disasters to the deep-sea ecosystem.

The main question regarding potentially polluting wrecks is, of course, what is the potential for a major spill? This subject has been written about extensively and excellent discussions are presented by Michel et al. (2005) and Landquist et al. (2013). While there is no clear consensus on the best comprehensive approach for risk remediation between coastal states who have dealt with the issue, a collective approach to the problem is incident specific and requires evaluation of the wreck site, environmental conditions, methods of risk reduction, and potential for catastrophic pollutant release versus slow leaks (Landquist et al., 2013: 91; Etkin et al., 2009). This is particularly true for older historic shipwrecks. 'It is clear that most of the oil remaining on these wrecks will eventually be released. More than 75 percent of the wrecks date back to World War II and thus have been underwater for 55–65 years, so there is added concern that corrosion will lead to increased oil discharges' (Michel et al., 2005: iv).

The sheer number of losses of wrecks that both polluted at the time of their loss and those which retain the potential to pollute is extensive. Germany's U-boat campaign, the 'Battle of the Atlantic,' sank some 3500 merchant ships, 175 Allied warships, and lost 765 U-boats in European, North Atlantic, Gulf of Mexico and Caribbean waters (Delgado, 2019: 337–338). Total allied warship losses in World War II, globally, neared 2000 vessels. In the Pacific and Indian Oceans, allied forces sank 686 Japanese warships, and 2346 merchant vessels during World War II (Joint Army-Navy Assessment Committee, 1947). As the corrosion of thousands of these steel hulls in the marine environment persists, the risk of accelerated leaking or catastrophic release increases with time in nearly every ocean on the planet as a result of that global conflict.

The determination of a shipwreck as environmentally hazardous or potentially polluting does not, however, negate its significance as a historic site. While the pollution risk from PPWs can pose a threat to ocean heritage in the environmental sense, the potential hazard also poses a risk to the archaeological integrity of the historic sites. NOAA's Remediation of Underwater Legacy Environmental Threats (RULET) project was designed to assist 'in prioritizing potential threats to ecological and socio-economic resources while at the same time assessing the historical and cultural significance of these nonrenewable cultural resources' for shipwrecks in US waters (NOAA, 2013: 1). Consistent with the National Historic Preservation

Act, RULET not only assessed which shipwrecks in US waters might pose a risk of polluting, but also addressed whether they met the standard for heritage listing.

Thankfully, not only for heritage reasons but also because some of these wrecks contain human remains, they are rightly seen as war grave sites, calling for respectful treatment even under challenging circumstances. The USS *Arizona* provides a good example of one approach, a memorial park in which there has been no intervention to remove pollutants, but the National Park Service periodically cleans up the sheen (see Glover, Chap. 4, this volume). Another example is the UK decision to remove the oil from the HMS R*oyal Oak* after an environmental assessment and long deliberations on the desire not to disturb the war grave, unless absolutely necessary to address the threat to the marine environment, and the livelihoods dependent upon a clean ocean (see Hill et al., Chap. 6, this volume).

Similarly, at the site of USS *Mississinewa* at Ulithi Atoll, typical approaches to oil removal have involved in depth assessments of the wreck sites' condition, stability, and integrity of the hulls, which enables archaeological documentation and characterisation. Methods of tapping hulls at locations over cargo tanks and pumping petroleum cargoes do not usually compromise the stability of the structure and rarely impact the sites' archaeological integrity or disturb any human remains should they exist within the hull. In determining how to remediate pollution risks, as each site is a case-by-case matter dependent on environmental conditions and site parameters, however, the potential environmental hazard must avoid or minimize harm to the integrity of the historic site. The *in situ* preservation policy that prefers non-intrusive activities is in harmony with the respectful treatment of these wrecks that are also war graves.

While a shipwreck site comes into equilibrium over time with the marine environment, corrosion is a constant process and steel hulled ships will eventually rust away; at some point throughout the lifecycle of an oil tanker shipwreck, there will be a release of its cargo. This is an example of an extracting filter, which removed elements of a ship from the site during its formation process, as defined by Muckelroy (1978). Bottom fishing gear has been shown to severely impact wrecks and has likely caused oil release events due to strikes (e.g., Brennan, 2016; Delgado et al., 2018; Brennan et al., 2023), and is a more event-based extracting filter compared to corrosion, which is constant. While shipwrecks in deep water are out of reach of such anthropogenic impacts such as trawl and dredge strikes, corrosion remains a factor that will in time cause petroleum cargoes to release. And, with the advent of deep-sea mining, even PPWs in areas under the high seas are at risk of an inadvertent spill, harm to the historic wreck and possible disturbance of a war grave.

The imperative factor in assessing and mitigating pollution risk from PPW sites, and in characterising their historic and archaeological integrity, is accessing them, a process that begins with locating unfound wrecks and being able to get divers or ROVs on site to systematically document them. Larger scale ocean mapping and exploration is needed to find many expected PPWs that lie in deep water. This can then be followed by full assessment and characterisation of the sites as an initial baseline from which further corrosion and hull integrity can be assessed in the future. In the modern era, we have better technology for surveying and accessing the deep sea, detecting shipwreck sites on the seabed, and post-*Deepwater Horizon*, better modeling for how oil behaves in the marine environment following a spill. The combination of modern technology and the time that has passed since World War II that left these wrecks on the seabed poses an increasing pollution risk that we have the means to address, if resources were devoted to proactive mitigation.

1.2   RUST, RULET, and the Potentially Polluting Wrecks Study

The Potentially Polluting Wrecks effort at NOAA for shipwrecks in US waters began with the compilation of a database known as Resources and UnderSea Threats (RUST) in 2002 by the Office of National Marine Sanctuaries following a series of mystery oil spills in sanctuaries waters, and which led to the identification of oil from the wreck of *Jacob Luckenbach* on California beaches. The RUST database was designed to be incorporated into the Sanctuaries Hazards Incident Emergency Logistics Database System (SHIELDS) developed by NOAA's Office of Response and Restoration's Hazardous Materials Division. 'The mission of the RUST database is to develop, maintain, and manage an active and comprehensive inventory of undersea threats and potential environmental hazards within United States waters' (Overfield, 2004: 74). Following its establishment within the Office of National Marine Sanctuaries, RUST was expanded to include 'all marine waters in US coastal zone and EEZ' (Zelo et al., 2005: 807). This work built upon previous PPW work from the International Oil Spill Conference as reported by Michel et al. (2005).

A congressional appropriation in 2010 authorised the RUST database to be expanded and evaluated to identify the most significant environmental threats from shipwrecks in US waters. This led to the Remediation of Underwater Legacy Environmental Threats (RULET) project, which correlated physical, historical, and environmental data with the shipwrecks in RUST and analyzed those data to determine which wrecks posed pollution risks. As RUST was 'not an effective risk assessment tool in its original form [because] it did not address the larger question of how to characterise potential pollution threats,' (Symons et al., 2014: 784) additional data and analysis was needed. The RULET project narrowed down some 20,000 wrecks to 573 within the US EEZ that could pose a substantial pollution threat, and then further to 87 'potentially polluting wrecks' or PPWs based on modeling of potential oil release based on worst case and probable discharge (NOAA, 2013; Symons et al., 2014). The final report in 2013 was the Risk Assessment for Potentially Polluting Wrecks in U.S. Waters (NOAA, 2013) and was accompanied by a Screening Level Risk Assessment Package for each of the 87 shipwrecks. This project was conducted within NOAA's Office of National Marine Sanctuaries, working with the Office of Restoration and Response, NOAA's Maritime Heritage Program, and the U.S. Coast Guard.

In order to refine the list of shipwrecks in US waters that could pose an environmental risk, the RULET study used a series of risk factors and data to assess the potential structural integrity of the wrecks and the likely amounts of oil remaining on board, and factors that might influence removal operations. These include the amount of oil on board including both cargo and bunker fuel, if the wreck may have been demolished as a risk to navigation, if a large amount of oil was likely lost during the sinking, and the nature of the sinking event (NOAA, 2013: ES-3). Response recommendations, in turn, included a variety of options from leaving shipwrecks alone and responding to leaks or spills ad hoc, to full oil removal and salvage. Intermediate options include background research to develop wreck-specific response plans, site monitoring, and *in situ* assessments through remote sensing and ROV observation (NOAA, 2013: 74).

Factors affecting planning response options include the type and volume of oil on board, water depth and visibility, location relative to shore, and other site factors like the wreck orientation and condition. Importantly, the study addresses the potential impact of remediation on shipwrecks' historical integrity: 'this vessel is of historic significance and will require appropriate actions under the National Historic Preservation Act (NHPA) and the Sunken Military Craft Act (SMCA) prior to any actions that could impact the integrity of the vessel' (NOAA, 2013: 6). Such assessments are essential to protecting shipwrecks that are historically significant and war graves. However, in many of the cases in the NOAA PPW study, the wrecks had not yet been located, so the response options were left high-level and all encompassing.

Following the release of the study, three subsequent actions by NOAA further refined the study. The first was defining a process by which a legal determination of historical and archaeological significance, through the existing program of the US Government, the National Register of Historic Places. The second was, through opportunistic studies by NOAA and partners, conducting physical, non-disturbance assessment of some of the 87 high risk wrecks. The first was the freighter SS *Fernstream*, sunk in a collision in 1952 near San Francisco Bay's Golden Gate Bridge; the second was the SS *Coast Trader*, sunk off the coast of British Columbia in 1942. The third was the USNS *Mission San Miguel*, a World War II era naval tanker lost at Maro Reef in the Northwestern Hawaiian Islands in 1957 (see Delgado, Chap. 5, this volume. As a result of those assessments, the potential for risk was downgraded for all three wrecks. Concurrently, NOAA's Office of National Marine Sanctuaries, working with partners, was also conducting a detailed survey and assessment of Battle of the Atlantic losses off the coast of North Carolina as part of a proposed expansion of USS *Monitor* National Marine Sanctuary between 2008 and 2014. These detailed assessments of wrecks, some of them listed in the PPW study but not determined to be high-risk, nonetheless were key in developing the right level of documentation for World War II PPW sites for heritage assessment, especially four German U-boats, two Allied warships, and several merchant ship losses in shallow and deep water (Delgado, 2019: 339–342). The RUST/RULET and Potentially Polluting Wrecks databases and studies produced an important assessment of the wrecks in US waters and helped compile information on the wrecks, especially those yet to be found, as a vital tool in beginning the efforts to locate them.

1.3   Site Assessment and Pollution Risk Remediation

The waging of two world wars over the course of three decades of the early twentieth century left oil-filled hulls on the bottom of the sea across the entire globe. Aside from those in US waters, high concentrations of potentially polluting wrecks are in the waters surrounding the United Kingdom, the Baltic Sea, and a large swath of the South Pacific. As these hulls corrode and begin to leak, coastal states have had to determine how best to remediate and mitigate these pollution risks. The International Oil Spill Conference study reflected on the growing demand for proactive oil removal from wrecks containing petroleum cargoes rather than the common reactive approach to address the environmental hazards once oil had begun to leak (Michel et al., 2005: 1). Like NOAA's approach, Michel and colleagues addressed the multifaceted approach required to assess and mitigate oil spill risk from wrecks, including spill modeling, costs and technology for diver versus ROV work, vessel needs, oil disposal, and the mechanics of removing heavy fuel oil in colder deep water. This global assessment also illustrated the significant spike in potentially polluting wrecks sunk during World War II, defined as those sunk between 60–70 years ago, in steel hulls corroding ever since in the ocean (Michel et al., 2005: 17); and nearly 20 years have passed since this study. As this study concludes, 'It is clear that most of the oil remaining on these wrecks will eventually be released' (Michel et al., 2005: iv).

Expanding efforts and funding for ocean exploration to locate and assess potentially polluting wrecks is an overarching need to investigate the myriad factors that impact a shipwreck site's condition, ability to retain oil cargo, and ultimately release it. Liddell and Skelhorn (2018: 83) list the important factors to assess as the type of wreck (i.e., warship or merchant vessel), the nature of the sinking event (the number of torpedo strikes, for instance, or how intact might the hull be), natural factors such as corrosion, and anthropogenic factors such as trawling and salvage. The UK government and Ministry of Defence (MOD) have dealt with the subject of polluting wrecks extensively due to having one of the largest inventories of state-owned wrecks in the world due to its large maritime role in both world wars, as well as the UK's policy of non-abandonment (Liddell & Skelhorn, 2018: 84). Leaking oil from the World War I battleship wreck of HMS *Royal Oak* in Scapa Flow led the UK to move from a response policy to one of proactive risk management and the management of the UK inventory of wrecks by Salvage and Marine Operations (SALMO) starting in 2008. This approach of assessing wreck sites and mitigating pollution risk proactively has also recently been done by the United States, as previously noted, with the remediation of the wrecks of USS *Mississinewa* at Ulithi Atoll in 2003, USS *Chehalis* in Pago Pago harbor, American Samoa, the German cruiser *Prinz Eugen* at Kwajalein (Naval Sea Systems Command, 2003, 2011, 2019), and the oil tankers *Coimbra* and *Munger T. Ball* off Long Island and Key West (Brennan et al., 2023).

The proactive approach to remediating chronically leaking oil tanker wrecks was recently conducted by the US Coast Guard and Resolve Marine Group on two shipwrecks on NOAA's PPW list, *Coimbra* and *Munger T. Ball*. Both wrecks were known to have been leaking for decades and therefore high priority targets for remediation. Tapping of the oil tanks through the mostly capsized hulls by divers succeeded in removing much of the petroleum product on board the wrecks. Also important to the efficacy of on-site wreck assessments was that these two projects had the author on site for the ROV and diver assessments, which was the first such projects in U.S. waters to have a maritime archaeologist as part of the on-board team. This work also succeeded in correcting a misidentification of the *Munger T. Ball* wreck as that of tanker *Joseph M. Cudahy*, which was sunk on the same day in 1942 by U-507, but which had different cargo capacities and carried different fuel types (Brennan et al., 2023).

Prioritisation of PPW remediation operations is another difficult and complex task. Understanding the potential hazards and prioritising which wrecks pose the greatest risk is paramount, as 'it is economically unfeasible and impractical to remediate all sunken shipwrecks' (Landquist et al., 2013: 86). A study by members of the British MOD and colleagues called for better standardisation of pollution risk assessment 'methodologies at both national and international levels' to develop more 'reliable means of prioritizing PPW' (Goodsir et al., 2019: 291). Such standard approaches could help in meeting goals set forth by a variety of conventions that call for protection of both Underwater Cultural Heritage (UCH) sites and marine environments, such as the UNESCO convention of 2001 and the Nairobi Convention of 2007, the latter of which additionally addresses the legal obligation for the removal of pollution hazards from the marine environment (Goodsir et al., 2019; see Aznar and Varmer, Chap. 2, this volume). The staged approach put forth for PPW site assessments is similar to that developed by NOAA and includes environmental desk-based assessment (E-DBA) followed by on-site wreck integrity evaluation along with environmental characterisation, followed by intervention and remediation as the final stage when determined necessary (Goodsir et al., 2019). Proposed responses included to both chronic and acute leak scenarios taking place over continual or catastrophic leak events and acknowledge the political pressures that are often applied to oil spill incidents, but which can be avoided through such proactive approaches.

The work presented by Carter and colleagues follows on previous prioritisation work for PPWs and presents an approach for the span of World War II wrecks in the Pacific. While the removal of petroleum product from PPW sites is sometimes undertaken, the initial desktop assessment and prioritisation is conducted irrespective of that remediation action; the desktop and then on-site assessments are used to assist in future remediations, but on the first level, 'help prioritize efforts towards more effective management and allocation of resources' (Carter et al., 2021: 7). This was conducted on the numerous World War II wrecks sunk at Chuuk lagoon, which, ahead of any remediation efforts, have identified 17 of the wrecks as the highest risk, for example, *Rio de Janeiro Maru*, a Japanese auxiliary ship sunk during Operation Hailstone in 1944. On-site assessments can also take into account factors in addition to cargo, vessel type, and depth, such as oceanographic parameters and the levels of corrosion present on the hull (Carter et al., 2021). Corrosion in the marine environment is the main driver of accelerated pollution risk as the steel hulls containing fuel and liquid cargoes deteriorate (e.g., Macleod, 2002, Glover, Chap. 4, this volume). Diver and ROV inspection of the hull of *Coimbra* in 2019, for example, found that the leaks were primarily coming from corroded rivets (Brennan et al., 2023).

1.4   Protecting Our Ocean Heritage

Finding, assessing, and potentially mitigating the shipwreck sites of potentially polluting wrecks is an imperative step toward managing and protecting our ocean heritage. The subject of Potentially Polluting Wrecks, and the topic of this volume, is central to the need to protect ocean heritage; PPWs are both historic sites of cultural significance and pose a threat to the marine environment, a juxtaposition that complicates protecting both cultural resource and marine ecosystems. Ocean heritage is both cultural and natural heritage in the material culture from past human activity and humanity's use of the maritime landscapes throughout history to the present day and into the future. Addressing the risk of pollution events into the marine ecosystem from PPWs works toward protecting the marine environment from further anthropogenic spill events while also documenting and assessing the wrecks as historic sites and, in many cases, war graves. As previously stated, the imperative step toward addressing PPWs is increased ocean exploration and mapping to locate undiscovered sunken ships that may pose a pollution risk so that steps toward assessing and potentially mitigating those risks can be undertaken, along with archaeological documentation of the wrecks for preservation and chronicling the wrecks' histories.

The chapters that follow will cover the legal context, environmental impacts, archaeology, and case studies of PPWs in certain areas of the Atlantic, Pacific, and the Baltic Sea. The book does not have case studies in all the areas where PPWs exist, and future work to address pollution risks will expand the geographic scope of the subject. Please contact The Ocean Foundation if you are interested in doing education and outreach on PPWs.

References

Brennan, M. L. (2016). Quantifying impacts of trawling to shipwrecks. In M. E. Keith (Ed.), *Site formation processes of submerged shipwrecks* (pp. 157–178). University Press of Florida.
Brennan, M. L., Delgado, J. P., Jozsef, A., Marx, D. E., & Bierwagen, M. (2023). Site formation processes and pollution risk mitigation of World War II oil tanker shipwrecks: *Coimbra* and *Munger T. Ball. Journal of Maritime Archaeology, 18*, 321–335.

Carter, M., Goodsir, F., Cundall, P., Devlin, M., Fuller, S., Jeffery, B., & Talouli, A. (2021). Ticking ecological time bombs: Risk characterisation and management of oil polluting World War II shipwrecks in the Pacific Ocean. *Marine Pollution Bulletin, 164*, 112087.

Delgado, J. P. (2019). *War at sea: A shipwrecked history from antiquity to the twentieth century.* Oxford University Press.

Delgado, J. P., Cantelas, F., Symons, L. C., Brennan, M. L., Sanders, R., Reger, E., Bergondo, D., Johnson, D. L., Marc, J., Schwemmer, R. V., Edgar, L., & MacLeod, D. (2018). Telepresence-enabled archaeological survey and identification of SS Coast Trader, Straits of Juan de Fuca, British Colombia, Canada. *Deep-Sea Research Part II, 150*, 22–29.

Etkin, D. S., van Rooij, H., & McCay French, D. (2009). *Risk assessment modeling approach for the prioritization of oil removal operations from sunken wrecks.* Effects of Oil on Wildlife.

Goodsir, F., Lonsdale, J. A., Mitchell, P. J., Suehring, R., Farcas, A., Whomersley, P., Brant, J. L., Clarke, C., Kirby, M. F., Skelhorn, M., & Hill, P. G. (2019). A standardised approach to the environmental risk assessment of potentially polluting wrecks. *Marine Pollution Bulletin, 142*, 290–302.

International Union for Conservation of Nature (IUCN). (2023). Marine Pollution from Sunken Vessels. Issues Brief, April 2023. https://www.iucn.org/resources/issues-brief/marine-pollution-sunken-vessels

Joint Army-Navy Assessment Committee (NAVEXOS P-468). (1947). Japanese naval and merchant shipping losses during world war II by all causes. Washington, DC.

Landquist, H., Hassellöv, I.-M., Rosén, L., Lindgren, J. F., & Dahllöf, I. (2013). Evaluating the needs of risk assessment methods of potentially polluting shipwrecks. *Journal of Environmental Management, 119*, 85–92.

Liddell, A., & Skelhorn, M. (2018). In K. Bell (Ed.), *Deriving archaeological information from potentially-polluting wrecks. Bridging the gap in maritime archaeology: Working with professional and public communities* (pp. 81–89). Archaeopress Publishing.

MacLeod, I. D. (2002). In situ corrosion measurement and management of Shipwreck Sites. In *International handbook of underwater archaeology* (pp. 697–714). Kluwer Acaemic/Plenum Publishers.

Michel, J., Gilbert, T., Waldron, J., Blocksidge, C., Etkin, D. S., & Urban, R. (2005). Potentially polluting wrecks in marine waters. In *2005 international oil spill conference, IOSC 2005*.

Muckelroy, K. (1978). *Maritime archaeology.* Cambridge University Press.

Naval Sea Systems Command. (2003). U.S. Navy salvage report, USS *Mississinewa* oil removal operations. Naval Sea Systems Command, S0300-B6-RPT-010, 0910-LP-102-8809.

Naval Sea Systems Command. (2011). U.S. Navy salvage report, ex-USS *Chehalis* fuel removal operations. Naval Sea Systems Command, S0300-B7-RPT-010, 0910-LP-110-9528.

Naval Sea Systems Command. (2019). U.S. navy salvage report, ex-USS *Prinz Eugen* oil removal operations. Naval Sea Systems Command, S1740-AK-RPT-010, 0910-LP-118-8301.

NOAA. (2013). Risk assessment for potentially polluting wrecks in U.S. Waters. https://nmssanctuaries.blob.core.windows.net/sanctuaries-prod/media/archive/protect/ppw/pdfs/2013_potentiallypollutingwrecks.pdf. Accessed 6 Apr 2022.

Overfield, M. L. (2004). Resources and UnderSea Threats (RUST) database: An assessment tool for identifying and evaluation submerged hazards within the National Marine Sanctuaries. *Marine Technology Society Journal, 38*(3), 72–77.

Symons, L., Michel, J., Delgado, J., Reich, D., McCay, D.F., Etkin, D.S., Helton, D. (2014). The Remediation of Underwater Legacy Environmental Threats (RULET) risk assessment for potentially polluting shipwrecks in U.S. waters. In *International oil spill conference proceedings*, May 2014.

Zelo, I., Overfield, M., & Helton, D. (2005). NOAA's abandoned vessel program and resources and under sea threats project—Partnerships and progress for abandoned vessel management. *NOAA Restoration and Response. International Oil Spill Conference Proceedings, 2005*(1), 807–808.

Open Access This chapter is licensed under the terms of the Creative Commons Attribution 4.0 International License (http://creativecommons.org/licenses/by/4.0/), which permits use, sharing, adaptation, distribution and reproduction in any medium or format, as long as you give appropriate credit to the original author(s) and the source, provide a link to the Creative Commons license and indicate if changes were made.

The images or other third party material in this chapter are included in the chapter's Creative Commons license, unless indicated otherwise in a credit line to the material. If material is not included in the chapter's Creative Commons license and your intended use is not permitted by statutory regulation or exceeds the permitted use, you will need to obtain permission directly from the copyright holder.

Chapter 2
Potentially Polluting Wrecks and the Legal Duty to Protect Our Ocean Heritage

Mariano J. Aznar and Ole Varmer

2.1 Introduction

As said in the introduction to this book, the determination of a shipwreck as environmentally hazardous or potentially polluting does not negate its significance as a historic site (Brennan, Chap. 1, this volume). Practice shows how wrecks may be different things at the same time: a lost property, an artificial reef, a gravesite, an obstacle for navigation, or/and a historic site and a potentially polluting wreck. This makes them a complex object for regulation, both at the domestic and international level (Aznar, 2015).

The purpose of this chapter is to elucidate part of the legal framework for cultural heritage that may also pose a threat to the environment. The last few decades have witnessed a longstanding interest in the protection of the marine environment, which today is generally codified in different international treaties, both on the law of the sea and environmental law. However, the other component of Ocean Heritage—the cultural, archaeological, or historic one—to date has received much less attention. Only in 2001, with the adoption of the UNESCO Convention of the

M. J. Aznar (✉)
University Jaume I, Castellón de la Plana, Spain
e-mail: maznar@uji.es

O. Varmer
The Ocean Foundation, Washington, DC, USA

© The Author(s) 2024
M. L. Brennan (ed.), *Threats to Our Ocean Heritage: Potentially Polluting Wrecks*, SpringerBriefs in Underwater Archaeology,
https://doi.org/10.1007/978-3-031-57960-8_2

Protection of the Underwater Cultural Heritage (2001 Convention),[1] was a more encompassing approach to both cultural and natural heritage adopted. The definition of underwater cultural heritage (UCH) given in this Convention always links both archaeological and natural context,[2] and all activities directed to UCH must include an environmental policy adequate to ensure that the seabed and marine life are not unduly disturbed. There is thus an intimate link between natural and cultural heritage, which may be traced back to 1972 when two seminal documents were adopted: the 1972 Stockholm Declaration, which introduced the precautionary approach in the international environmental agenda, and the 1972 World Heritage Convention (WHC), where the wisdom of integrating the conservation of both natural and cultural heritage became international law. A decade later, the 1982 Law of the Sea Convention (LOSC) included the duties to protect the marine environment and objects of an archaeological and historical nature. Finally, the 2001 Convention crystallised such twofold duty. The 1972 WHC, 1982 LOSC, and 2001 Convention thus provide the general legal foundation and guidance for conserving our Ocean Heritage.

The definition of UCH given in this Convention establishes a time-limit of 100 hundred years, which is conservative and may be lower in the implementing domestic law. Sunk in 1912, the wreck of RMS *Titanic* is today covered by the 2001 Convention, as are the wrecks of vessels sunk during World War I (1914–1918). In less than two decades, thousands of PPWs sunk during World War II (1939–1945) will be also covered by the 2001 Convention as UCH. Under the domestic law of many nations, WWII wrecks are already considered historic. For example, the Australian UCH Act provides blanket protection for wrecks underwater for at least 75 years. The United States' National Register of Historic Places uses 50 years as a rule of thumb for historic properties and eligibility for protective procedures. As many PPWs are sunken state craft of warring nations, this chapter touches on the sensitive issues of ownership, sovereign immunity, and the fact that many sites are also wartime graves.

How to preserve historic PPWs from a cultural perspective while monitoring, mitigating or even removing them as threats to the marine environment is the legal

---

[1] During negotiations some countries led by the United States expressed concern that the instrument creates new rights within the EEZ/continental shelf beyond those recognised in the LOSC and thus maybe upsetting the balance of interests under the 1982 LOSC. There was also some concern about whether the consent of the foreign flagged State before the authorisation of activities directed as sunken warships within the territorial sea as was clear in the other maritime zones. However, the international community, including the US, recognise the coastal State authority and jurisdiction to address threats to the marine environment within the EEZ including sunken warships that may also be in the territorial sea.

[2] UCH is defined as 'all traces of human existence having a cultural, historical or archaeological character which have been partially or totally under water, periodically or continuously, for at least 100 years such as: (i) sites, structures, buildings, artefacts and human remains, together with their archaeological *and natural context*; (ii) vessels, aircraft, other vehicles or any part thereof, their cargo or other contents, together with their archaeological *and natural context*; and (iii) objects of prehistoric character' (emphasis added).

challenge. There are much older shipwrecks that may pose an environmental risk. For example, the 1724 wreck of *Tolosa*, a Spanish galleon sunk off the Dominican Republic, then carrying tons of mercury. However, the main concerns are generated by PPWs sunk during the last century, most in World War I and II.

For this, considering the duty to protect our Ocean Heritage and the precautionary principle nested in international and domestic law from the 1972 Stockholm Declaration, the next pages will analyse how they influenced the WHC, LOSC and some International Maritime Organisation (IMO) conventions. We will see how the discussion of the 1992 Rio Declaration built upon the 1972 Declaration, including the integrated management of natural and cultural heritage, and taking a precautionary approach to activities in balancing economic development with the conservation of our Ocean Heritage for future generations. As mentioned, the 2001 Convention echoes all these principles; and today the UN Decade of Ocean Science for Sustainable Development builds upon this rich history and is relevant to the challenges in addressing the threat to our Ocean Heritage from PPWs. Addressing the threats to our Ocean Heritage from PPWs comes within these outcomes sought by the Decade: a clean ocean where sources of pollution are identified and reduced or removed; a healthy and resilient ocean where marine ecosystems are understood, protected, restored, and managed; and a productive ocean supporting sustainable food supply and a sustainable ocean economy. Many of the PPWs are within the waters of lesser developed nations where the livelihoods of their peoples are dependent upon a healthy ocean and coastal waters for fishing and ecocultural tourism.

2.2 The Initial Evolution of a Duty to Protect Our Ocean Heritage

2.2.1 1972 Stockholm Declaration

The United Nations Conference on the Human Environment met in Stockholm and resulted in the Stockholm Declaration. It codified in programmatic terms the customary practice of nations in balancing economic development and protecting the environment so that it may be inherited by future generations in a healthy state (heritage). It contains principles that document and delineate the duty to protect and to cooperate for that purpose under customary international law, including that we bear 'a solemn responsibility to protect and improve the environment for present and future generations' (Principle 1). A 'special responsibility to safeguard and wisely manage the heritage of wildlife and its habitat, which are now gravely imperiled by a combination of adverse factors. Nature conservation, including wildlife, must therefore receive importance in planning for economic development' (Principle 4). 'States should adopt an integrated and coordinated approach to their development planning to ensure that development is compatible with the need to protect and improve environment for the benefit of their population' (Principle 13). 'In exercising the sovereign right to exploit their own resources, there is the responsibility to ensure that activities within their jurisdiction or control do not cause damage to the environment of other States or of areas beyond the limits of national jurisdiction' (Principle 21).

The 'integrated and coordinated approach' to development planning in Principle 4 helps to delineate the precautionary approach, being the starting point for the introduction of concepts into international law that previously were only used in national legislation. This is reinforced by the integration of natural and cultural resources to be conserved in UNESCO's 1972 World Heritage Convention.

2.2.2 The World Heritage Convention Integrating the Conservation of Cultural and Natural Heritage

While the focus in Stockholm was on the environment and sustainable development, at UNESCO in Paris the focus was an agreement to protect cultural and natural heritage. With 193 State parties today, the World Heritage Convention remains one of the most widely accepted treaties. While it started with recognising terrestrial sites and traditional cultural structures, inclusive of coastal sites with marine components (mostly involving sites in Europe, the Americas and Africa), it more recently evolved to recognise heritage in the marine environment even beyond the territorial sea into the exclusive economic zones (EEZ) and continental shelves (CS), particularly in the Asia–Pacific marine environment. For example, the Papahānaumokuākea Marine National Monument in Hawai'i was inscribed on the World Heritage 'mixed list' for its 'outstanding universal value' as both a natural and cultural heritage site. Of note, is that some of the natural heritage, like coral, were also recognised as cultural heritage of the native Hawaiian people and some of their practices were recognised as intangible cultural heritage.[3] This move seaward continues as there are now calls for recognition of heritage in the high seas, including wreck sites such as *Titanic*. The challenges involve ensuring consistency with the LOSC, more specifically, implementing the duty to protect and cooperate through reliance of flag State jurisdiction, and Port State jurisdiction for enforcement of activities directed at cultural heritage under the high seas.

However, the Convention does not formally apply to EEZ/CS of States and to areas beyond national jurisdiction (ABNJ) like the High Seas and the 'Area,'[4] that include rich marine environment, both cultural and natural. This is why the Report 'World Heritage in the High Seas: An Idea Whose Time Has Come' (UNESCO, 2016) proposes a strong movement towards the applicability of the WHC beyond national jurisdiction.

---

[3] See more information at WHC site: https://whc.unesco.org/en/list/1326

[4] The 'Area' is defined in Art. 1(1)(1) LOSC as 'the seabed and ocean floor and subsoil thereof, beyond the limits of national jurisdiction.'

2.3   The 1982 Law of the Sea Convention: The General Legal Framework for Managing PPWs

Marine environmental questions were discussed during the First and Second UN Conferences on the Law of the Sea. But it was during the Third Conference (UNCLOS III) when the marine environment was at the heart of many discussions when delineating the future UN Convention on the Law of the Sea (LOSC), labelled as the 'Constitution of the Sea', adopted in 1982 and entered into force in 1994.

The LOSC is well recognised as a codification and progressive development of the law of the sea. It balances the flag State rights of navigation, fishing, marine research, mining and other uses with the coastal State jurisdiction and authority in various maritime zones including a 12 nm territorial sea, a 24 nm contiguous zone, a 200 nm exclusive economic zone (EEZ), the continental shelf, the high seas, and the Area.

LOSC provides the general legal framework for the use and protection of the marine environment, including natural and cultural heritage resources. This legal framework includes several articles on the duty to protect the marine environment but only a couple on protecting 'objects of an archaeological and historical nature.'

2.3.1   General Provisions on the Duty to Protect the Marine Environment and PPWs

LOSC has provisions protecting marine environment in the different marine zones. For example, on the conservation of living resources in the EEZ (Art. 61) or in the High Seas (Arts. 116–120), or a general environmental duty in Art. 145 for the Area. But it is in its Part XII (Arts. 192–237) where it is found the core duties on marine environment protection.

Under Articles 192 and 194 there is the general obligation of States to protect and preserve the marine environment. This general principle may be applied both to wrecks that become artificial reefs, deserving protection as part of the marine environment, and to wrecks that are potentially (or actually) endangering this environment because of their deteriorating structure or polluting cargo still aboard. Specifically, Art. 194(3)(b) deals with the measures to be taken by States, including those designed to fully minimise 'pollution from vessels, in particular measures for preventing accidents and dealing with emergencies, ensuring the safety of operations at sea, preventing intentional and unintentional discharges, and regulating the design, construction, equipment, operation and manning of vessels.' These general duties, as it will be seen, have been developed and completed by the different IMO conventions preventing and mitigating pollution from vessels.

2.3.2   Articles Protecting Cultural Heritage Found at Sea (303) and in the Area (149)

Cultural heritage was not prominent enough during UNCLOS III. There were some discussions towards the end of negotiations that resulted in only two articles: article 303, which is applicable to all maritime zones, and article 149 for heritage in the Area. However, they did not provide clear guidance on how to implement the obligations much less to address the threats to the marine environment posed by wrecks that may also be objects of cultural heritage.

Article 303(1) recalls the general obligation whereby 'States have the duty to protect objects of an archaeological and historical nature found at sea and shall cooperate for this purpose.' It can be argued that this general duty forms part of customary law, as evidenced by the practice of nations including non-parties like the US, Türkiye, Cambodia, or Colombia (Varmer, 2020). Paragraph 2 of Article 303, which has been declared as customary law (*Nicaragua/Colombia* case, International Court of Justice, 2022), is to be read today as granting coastal States 'the power of control with respect to archaeological and historical objects found within the contiguous zone', which goes beyond what article 303(2) explicitly says (Aznar, 2014). This means that any activity directed to a PPW considered UCH located in the contiguous zone needs to be authorised and regulated by the coastal State, thus applying not only its cultural heritage national legislation but its environmental legislation as well.

Beyond the outer limit of that contiguous zone (24 nm), LOSC left a perceived gap that related to the EEZs and continental shelves. Except to the duty to protect under Art. 303 (1), there is only a contextual and analogic interpretation of natural environmental rules that could apply to PPWs and perhaps protect cultural heritage in its natural context.

Beyond the outer limit of these two zones, for the Area article 149 provides that '[a]ll objects of an archaeological and historical nature found in the Area shall be preserved or disposed of for the benefit of mankind, particular regard being paid to the preferential rights of the State or country of origin, or the State of cultural origin, or the State of historical and archaeological origin.' While it is not always clear which nations have preferential rights, regarding the threat posed by PPW considered cultural heritage, it should be relevant to identify those nations with an interest and responsibility on these wrecks, which would include at least the flag States of the sunken ship and nation from which the cargo came (Aznar, 2019).

2.4   Maritime Law Conventions and the Ocean Heritage: The IMO Endeavours

2.4.1   Natural Disasters by Human Activities

Increasingly during the last century, maritime commerce has included hazardous cargoes aboard which may produce environmental disasters by accident, negligence, or fault. When the *Torrey Canyon* tanker spilled tons of oil in 1967, severely contaminating the marine environment and killing tens of thousands of living resources, it was a catalyst for the modern environmental movement, and nations recognised the need for coastal States to take measures to address pollution from outside their territory, including the high seas.

Gathered at the IMO, States and maritime operators realised the need to adopt new rules to avoid or minimise these disasters. Some of them should apply to PPWs, as for example the 1969 International Convention Relating to Intervention on the High Seas in Cases of Oil Pollution Casualties, asking Parties to prevent, mitigate or eliminate grave and imminent danger to their coastline or related interests from pollution or threat of pollution of the sea by oil, following upon a maritime casualty (Art. I). Or the 1972 Convention on the Prevention of Marine Pollution by Dumping of Wastes and Other Matter, which asks Parties to prevent the pollution of the sea by the dumping of waste and other matter that is liable to create hazards to human health, to harm living resources and marine life (Art. I), understanding dumping as including any deliberate disposal of vessels, aircraft, platforms or other man-made structures at sea (Art. III(a)). Thus, the creation of an artificial reef by the disposal of a vessel is acceptable, provided that such placement is not contrary to the aims of this Convention, the protection of marine environment (Art. III(b)(ii)). However, in both cases, the application to sunken warships (typically PPWs) is limited when not excluded.

2.4.2   Cultural Disasters by Human Greed

By the end of the 1970s, when scuba diving was technically possible and new underwater technologies were available, cultural heritage in oceans was also targeted by treasure hunters. US courts, sitting in Admiralty jurisdiction, awarded the treasure, applying what was called 'historic salvage' in the absence of clear application of historic preservation law (Varmer & Blanco, 2018). The 1989 London Salvage Convention developed at the IMO to incorporate rewards for salvors preventing or minimizing damage to the marine environment in their salvage of wrecks. While there is a provision for parties to declare that the Convention shall not be applied to historic wrecks, the salvage of our UCH continues. However, several admiralty decisions regarding historic sunken vessels, both State owned (*Juno* and *La Galga, Mercedes*) or private vessels (*Titanic*) have reversed the previous approach to 'historic salvage', now preserving UCH under strict conditions following archaeological standards widely accepted by the international community, which also include the protection of UCH in its natural context thus preserving our Ocean Heritage.[5]

2.4.3   The Wreck Removal Convention

The IMO conventions did not completely cover the duty to prevent or mitigate threats posed by PPWs. It has been estimated that there are three million wrecks worldwide, thousands of which are PPWs.

The 2007 IMO International Convention on the Removal of Wrecks, also known as the Nairobi Wreck Removal Convention, entered into force on April 14, 2015 and has 67 parties as of January 2024. The Convention recognises the right of a coastal State to address threats from foreign flagged wrecks that may have the potential to adversely affect the safety of lives, goods, and property at sea, as well as the marine environment, pose a hazard to the coastal State or to require the shipowner to remove the wreck at his own expense; hazard being defined as 'any condition or threat that: (a) poses a danger or impediment to navigation; or (b) may reasonably be expected to result in major harmful consequences to the marine environment, or damage to the coastline or related interests of one or more states.'

While the provisions and measures apply only when both the coastal State and the vessel's flag State are parties to the Convention, they may provide guidance for addressing the threats by non-parties and cases involving sunken warships. As is the case with many other maritime conventions, it does not apply to 'any warship or other ship owned or operated by a state and used, for the time being, only on Government non-commercial service' unless the flag State decides otherwise (Art. 4). The problem with the Wreck Removal Convention relies on its application only to wreckages produced after its entry into force (2015), which would leave out PPWs that originated during the two World Wars.

---

[5] In the US, this was possible thanks to a coordinated effort among different agencies, led by NOAA, which offered admiralty courts an acceptable set of conditions preserving the archaeological and natural context of historic wrecks.

2.5   A New Approach to Ocean Heritage

2.5.1   The Precautionary Approach as a Guide to Manage PPWs as UCH

Ten years after the LOSC and twenty years after the Stockholm conference, the United Nations convened another conference on environment and development in Rio de Janeiro, Brazil. This 1992 Rio Conference resulted in the Rio Declaration on Environment and Development (Rio Declaration), including Agenda 21 which provides that there is a duty to protect the marine environment and to cooperate for that purpose, expressly flowing from the LOSC. Chapter 17.1 of the Rio Declaration highlights how the LOSC 'sets forth rights and obligations of States and provides the international basis upon which to pursue the protection and sustainable development of the marine and coastal environment and its resources.' It then identifies approaches to implement this duty and specifically calls for integrated management and a precautionary approach, transformed into an international legal principle, in the sustainable development and protection of the marine environment. Principle 15 of the Rio Declaration reads as follows: 'In order to protect the environment, the precautionary approach shall be widely applied by States according to their capabilities. Where there are threats of serious or irreversible damage, lack of full scientific certainty shall be not used as a reason for postponing cost-effective measures to prevent environmental degradation.'

Since then, this principle has been incorporated into several international treaties, some of them addressing (albeit sometimes obliquely) PPWs; has been also included, for example, in the International Seabed Authority's Mining Code; and has been declared as part of general international environmental law, as may be seen in the case-law of the International Tribunal for the Law of the Sea (ITLOS). The International Seabed Authority (ISA) and UNCLOS States parties are negotiating a new regulation on exploitation of the mineral resources in the Area. There is an ongoing discussion on how to protect underwater cultural heritage in its natural context during deep-sea mining activities, which should be always presided by the precautionary approach.

As a recent landmark of this approach, the Agreement under the United Nations Convention on the Law of the Sea on the Conservation and Sustainable Use of Marine Biological Diversity of Areas beyond National Jurisdiction (BBNJ Treaty), adopted on 12 June 2023 but not yet in force, expressly includes the precautionary principle (Art. 5(d)), the need for environmental impact assessments (Part IV) and the governance and management of large portions of the oceans, even beyond national jurisdiction, using marine protected areas (Part III) under international monitoring (Art. 13).

2.5.2   2001 Convention on the Protection of Underwater Cultural Heritage

UNESCO convened a Meeting of Experts to negotiate an agreement to provide the sorely needed details to implement the duty to protect 'objects of an archaeological or historical nature' under the 1982 LOSC and to address the major threat to this heritage from treasure hunting. Consensus was reached quickly on a clear definition of UCH to address the undefined terms in the LOSC. Consensus was also reached on the general ban against the application of the law of finds and salvage with a narrow exception for nations that may want to implement the obligations under their domestic maritime law including the law of salvage (Varmer & Blanco, 2018). In that case, the implementation must be consistent with the entire Convention, including the Annex Rules which are the standards and requirements for when recovery or salvage is determined to be in the public interest.

The Convention has four main principles on which there was consensus: (1) the obligation to protect and preserve UCH; (2) the preferred first policy option of *in situ* preservation; (3) no 'commercial exploitation' of UCH; and (4) cooperation among States to protect UCH, particularly for training, education, and outreach. There was however a lack of consensus on the relation between LOSC and the Convention (with a fear of creeping jurisdiction of coastal States in the EEZ/CS), and on the legal status of sunken States vessels (particularly those located in the territorial sea). But all States agreed in the general precautionary approach embodied in the *in situ* rule, as the first option to preserve UCH before using more intrusive or destructive methodologies.

Hence, precaution in activities directed at UCH is particularly relevant if such heritage is a PPW. In this case, these activities must be performed under severe conditions keeping in mind that: (i) UCH is by definition inextricably linked to its natural context (Art.1(1)); (ii) any project must have an environmental policy to ensure that the seabed and marine life are not unduly disturbed (Rules 10 and 29, Annex); and (iii) as all preliminary work of the projected activity 'shall include an assessment that evaluates the significance and vulnerability of the underwater cultural heritage and the surrounding natural environment to damage by the proposed project' (Rule 14, Annex).

2.5.3   The Question of PPWs Which Are Both UCH and Warships or Other States Vessels

As mentioned, when drafting the 2001 Convention, consensus was not reached regarding the treatment by coastal States of foreign sunken State vessels within their Territorial Sea. Parties only agreed that 'consistent with State practice and international law, including [the LOSC], nothing in this Convention shall be interpreted as modifying the rules of international law and State practice pertaining to sovereign immunities, nor any State's rights with respect to its State vessels and aircraft.' Immunity and property status of State public vessels are clear in international law. Articles 32, 95 and 96 of LOSC reflect customary law, as it does article 16(2) of the 2005 UN Convention on Jurisdictional Immunities of States and Their Property. However, there is no international general convention on the immunities of *sunken* States vessels, particularly sunken warships if also PPW. Yet, the practice of nations regarding jurisdiction and control over PPWs is also informed by the 2015 Resolution of the Institut de Droit International on 'The Legal Regime of Wrecks of Warships and Other State-owned Ships in International Law.' (2015 Resolution).

Unless a sunken State vessel has been expressly abandoned, it continues to be owned by that flag State and is therefore subject to flag State jurisdiction (Articles 3–42,015 Resolution). Contemporary to the promulgation of the US Sunken Military Craft Act of 2004, France, Germany, Japan, Russia, Spain, and the UK made similar statements to that proclaimed by the US on immunity of sunken State vessels, including warships. However, this legal position should be balanced with the jurisdiction of coastal States in their different maritime zones.

The coastal State's sovereignty to regulate activities within its territorial waters is regardless of any foreign flag State ownership and immunity (Article 7, 2015 Resolution). In accordance with Article 303(2) of the LOSC, it may also regulate the removal of historic sunken State vessels from its contiguous zone (Article 8, 2015 Resolution) which would include requiring permits for the removal of oil, fuel, munitions, and other hazardous materials. Beyond the outer limit of the contiguous zone (EEZ/CS), general environmental rules as foreseen in the LOSC do apply, and the UNESCO 2001 Convention respects them in its Article 10(2) when expressing that "[a] State Party in whose exclusive economic zone or on whose continental shelf underwater cultural heritage is located has the right to prohibit or authorize any activity directed at such heritage to prevent interference with its sovereign rights or jurisdiction as provided for by international law including the United Nations Convention on the Law of the Sea." These sovereign rights or jurisdiction include coastal State's environmental rights. However, if such activity is directed to a historic PPW, the same article establishes in its paragraph 7 that "no activity directed at State vessels and aircraft shall be conducted without the agreement of the flag State [...]" Therefore, cooperation is important, in accordance with any other applicable treaties. Article 9 of the 2015 Resolution also recognizes that a coastal State has sovereign rights and jurisdiction to protect and manage the environment and resources of its continental shelf and EEZ, which includes addressing the threats from PPWs. This should be done in due regard to the rights of the foreign flag State of a PPW. Cooperation is important. However, if the flag State does not take any action after having been requested to cooperate, the coastal State may proceed and even remove the wreck.

As we have seen, sunken warships may be a polluting wreck and a historic resource or cultural heritage. They may also be a maritime war grave, deserving particular respect (Forrest, 2019). In these cases, the flag State has a particular interest and duty or responsibility regarding its sunken public vessels, which is shared with the coastal State if such PPW is sunk in or near its maritime zones. In any case, the activities, and duties should be always presided by the precautionary approach balancing the two intimate components of our Ocean Heritage: natural and cultural.

2.6   Conclusion

In light of the importance of conserving our Ocean Heritage for future generations, the legal duties to protect it under international law, and the goals of the UN Decade of Ocean Science for Sustainable Development, the best way to address the threats from PPWs is therefore a precautionary approach that involves a moratorium or pause against activities that may trigger these ticking time bombs until the proper science and assessments are done to make sure that these activities are truly sustainable. The moratorium would be temporary and limited to those activities directed at PPWs such as salvage. There should also be a temporary moratorium against certain indirect activities that could result in irreparable harm and destruction to UCH and marine life such as bottom trawling, or deep seabed mining until after surveys have been conducted to ensure that no PPWs are in planned exploitation areas, proper environmental impact assessments have been conducted, and significant natural and cultural heritage are set aside as marine protected areas.

References

Aznar, M. J. (2014). The contiguous zone as archaeological zone. *The International Journal of Marine and Coastal Law, 28*, 1–51.

Aznar, M. J. (2015). Regarding 'les épaves de navires en haute mer et le droit international. Le cas du Mont-Louis by Guido Starkle (1984/1985-I):' Sensitive wrecks, protecting them and protecting from them. *Revue belge de droit international, 1–2*, 32–46.

Aznar, M. J. (2019). The notions of 'preferential right' and 'interest' of sates in the protection of the underwater cultural heritage. *Revista Electrónica de Estudios Internacionales, 38*, 1–37.

Forrest, C. (2019). *Maritime legacies and the law. Effective legal governance of WWI wrecks.* E. Elgar Publishers.

ICJ. (2022). *Alleged violations of sovereign rights and maritime spaces in the Caribbean Sea (Nicaragua V. Colombia),* Judgement of 22 April 2022.

UNESCO. (2016). Freestone, D. Laffoley, D., Douvere, F. and Badman, T. *World heritage in the high seas: An idea whose time has come, see also world heritage Centre world heritage marine program.* http://whc.unesco.org/en/marine-programme.Hampi+World+Heritage+Site+Karn atakaHampi

Varmer, O. (2020). The duty to protect underwater cultural heritage and to cooperate for that purpose under law and policy. In H. Karan & K. Van Turk (Eds.), *The legal regime of underwater cultural heritage and marine scientific research* (pp. 77–117). University of Ankara.

Varmer, O., & Blanco, C. (2018). The case for using the law of salvage to preserve underwater cultural heritage: The integrated marriage of the law of salvage and historic preservation. *Journal of Maritime Law & Commerce, 49*, 401–424.

Open Access This chapter is licensed under the terms of the Creative Commons Attribution 4.0 International License (http://creativecommons.org/licenses/by/4.0/), which permits use, sharing, adaptation, distribution and reproduction in any medium or format, as long as you give appropriate credit to the original author(s) and the source, provide a link to the Creative Commons license and indicate if changes were made.

The images or other third party material in this chapter are included in the chapter's Creative Commons license, unless indicated otherwise in a credit line to the material. If material is not included in the chapter's Creative Commons license and your intended use is not permitted by statutory regulation or exceeds the permitted use, you will need to obtain permission directly from the copyright holder.

Chapter 3
Environmental Impact and Modeling of Petroleum Spills

Matt Horn, Deborah French-McCay, and Dagmar Schmidt Etkin

3.1 Introduction

The risks associated with a release of oil posed by potentially polluting wrecks (PPW) span a wide range of probabilities and potential magnitudes for environmental consequences. Even a lay reader will be well aware that following a release of oil, there is great potential for environmental damage and mortality of birds, mammals, and fish. Fisheries and beach closures and localised evacuations may occur to limit the exposure of humans to potential contaminants. The range of socio-economic and ecological impacts can be quite large between releases with the geographic extent and magnitude of effects being extremely variable between releases. In addition, the duration of these effects and changes to populations and ecosystems can range from a few days to years or even decades in some circumstances. This variability necessitates the quantitative assessment of the range of environmental impacts to understand where a release may occur, the environmental conditions at the time of the release, the geographic extent over which it may be transported, and the receptors of interest (e.g., species of concern, shorelines, populated areas) that may be impacted. Computational oil spill models were developed to characterise the movement and behavior of released oil in the environment, while also quantifying the duration of exposure to levels of contamination and their potential for both lethal and sublethal effects.

M. Horn (✉) · D. French-McCay
Tetra Tech, Inc., South Kingstown, RI, USA
e-mail: matt.horn@tetratech.com

D. S. Etkin
Environmental Research Consulting, Cortland Manor, NY, USA

© The Author(s) 2024   25
M. L. Brennan (ed.), *Threats to Our Ocean Heritage: Potentially Polluting Wrecks*, SpringerBriefs in Underwater Archaeology,
https://doi.org/10.1007/978-3-031-57960-8_3

3.2   Background

Countless ships have been lost at sea. A portion of these may contain petroleum products that have the potential to result in environmental impacts. The National Oceanic and Atmospheric Administration (NOAA) maintains a large database of shipwrecks, dumpsites, navigational obstructions, underwater archaeological sites, and other underwater cultural resources. This internal database, Resources and Undersea Threats (RUST), has over 30,000 targets, including approximately 20,000 shipwrecks in US waters (Overfield, 2004; Zelo et al., 2005). However, only a small fraction of these wrecks is likely to contain oil. Many vessels came to a violent end, breaking apart in storms, collisions, or in battle. Shallower wrecks were frequently salvaged or intentionally destroyed, as they posed risks to navigation. The remaining wrecks are therefore located at depth and have likely suffered from corrosion and the passage of time. However, reviews of historical information have identified thousands of sunken vessels along the U.S. coast, which have identified many with significant volumes of oil remaining on-board (French McCay et al., 2012, 2014; NOAA, 2013). In 2005, it was estimated that there are at least 8500 potentially polluting sunken wrecks of tankers and large vessels (at least 400 gross tons) worldwide (Michel et al., 2005). Unless located and remediated, these largely forgotten wrecks will begin to leak, often becoming the source of 'mystery spills' until the source is identified. Concerns lie with the potential for effects following a release with contaminants in the water column, on the water surface, and along nearby shorelines.

The RUST database (and those like it) is important from an environmental impact and modeling perspective as it provides risk assessors with a more complete understanding of where wrecks are located and what they likely contain. The range of locations enables the risk assessor to consider local environmental conditions (e.g., currents, tides, winds), key geographic features, and water depth, which will impact the potential transport of released oil and its potential to impact specific locations and/or receptors. As an example, an oil spill from a wreck that is located near the Gulf Stream off the coast of North Carolina has a greater potential for current transport, when compared to a wreck that may be in more stagnant waters and may affect amenity beaches along the coast. The range of product types and potential release volumes contained on each vessel enables the determination of trajectory and fate based upon known physical and chemical transport and fates processes. As an example, a large amount of marine diesel may evaporate, dissolve, and degrade quickly, and also be readily entrained in the water column, leading to a relatively small area for potential effects. A comparably smaller release of a heavier crude oil or fuel oil may be much more persistent and be transported greater distances, with a greater potential to impact beaches. The chemical composition of each oil will also impact its level of toxicity and potential to impact aquatic life. It is clear that there are numerous variables that must be considered when simulating the potential environmental consequences following a release.

3.3 Previous Work

The U.S. Coast Guard and the Regional Response Teams (RRTs), as well as NOAA, are responsible for the development of regional and area contingency plans. In 2010, the U.S. Congress appropriated $1 million to identify the most ecologically and economically significant potentially polluting wrecks (PPW) in U.S. Waters. NOAA worked closely with the U.S. Coast Guard Office of Marine Environmental Response Policy in implementing this mandate. The Remediation of Underwater Legacy Environmental Threats (RULET) effort supported the prioritisation of potential threats to coastal resources, while at the same time assessing the historical and cultural significance of these nonrenewable cultural resources (Symons et al., 2014). Similarly, there was an initiative using a slightly different screening approach to develop for the Mediterranean Sea with the Development of European Guideline for Potentially Polluting Shipwrecks (DEEPP), categorised with respect to Volume Class (estimated volume of oil on board) and Distance Class (proximity to shore) which are used to determine a risk factor (Alcaro et al., 2007). The RULET project narrowed the list of 20,000 vessels from RUST to 107 wrecks within the U.S. Exclusive Economic Zone (U.S. EEZ) that could pose a substantial pollution threat, which was further refined to 36 high priority worst case discharge (WCD) scenarios. From that, NOAA developed a total of 87 risk assessment packages for consideration by the U.S. Coast Guard Federal On-Scene Coordinators (FOSCs).

While each of the RUST, RULET, and DEEPP projects were sizeable investments, they operated under the premise that it would be impossible (from a financial and time perspective) to proactively respond to all wrecks immediately. Rather, the intent was to focus attention and limited resources on specific regions of interest that had the highest potential risk (likelihood and magnitude of potential consequences). Oil spill modeling was used to assess releases from numerous identified wrecks to begin to understand the range of potential consequences. The highest risk wrecks were then identified as focus points for oil spill response preparedness, surveillance, and potential mitigation efforts.

The components of an oil spill risk assessment for PPWs are essentially like that for other potential spill sources (Etkin et al., 2017; Etkin, 2019). Improvements have been made in the assessment of leakage probabilities, specifically in the VRAKA model (Landquist et al., 2014). However, given enough time, the probability of any specific wreck releasing oil will increase due to the continued corrosion of the hulls in the marine environment. The remainder of this chapter will focus on the assumption that oil has been released.

3.4   Released Oil

The environmental consequence following any release of oil depends on a number of factors, the two most obvious of which include the type (e.g., gasoline, diesel, crude oil, heavy fuel oil, etc.) and volume of oil. Oil is a complex mixture of many thousands of different hydrocarbon compounds derived from naturally occurring geological formations. Each of these compounds has its own molecular weight, density, viscosity, solubility, volatility, and toxicity. Therefore, the unique mixture of each oil has physical and chemical properties that reflect its unique composition, as well as its state of weathering from fresh oil along the continuum of weathering, which will affect its transport and fate, once discharged into the environment (NRC, 2003). In addition, these physical and chemical properties of each chemical compound within the oil impacts the ultimate toxicity of the oil, which will influence the magnitude of potential consequences following a release (e.g., mortality). The next most important factor influencing the potential consequence is the environmental conditions at the time of release for that specific release location. Environmental forcing parameters such as winds, currents, and tides will transport (i.e., advect) the oil, while other environmental parameters such as waves, temperature, and sunlight may change the dispersion, behavior, and/or weathering of the oil. Ultimately, the potential effects are dependent on both the concentration of contaminants and the duration of exposure to receptors (French-McCay, 2002) that may be sessile or mobile. Therefore, release rate of the product (amount of oil entering the environment over a period of time) will also play an important role in influencing the potential transport, fate, and resulting consequences.

To develop a robust understanding of the potential risks and the variability of effects from oil spills from PPWs, computational oil spill models may be used to assess the range of potential movement, behavior, and resulting consequences of a range of specific releases under a range of geographic locations and environmental release conditions. These tools can then be used to quantify the range of potential releases and resulting effects, which can be quite variable.

3.5   Computational Spill Models

Computational oil spill models such as OILMAP and SIMAP have been in development for over 40 years. The models are validated against real-world release and used extensively in the United States and internationally to meet regulatory requirements and to develop recommendations and guidelines. They are frequently used by industry, government, and academia. SIMAP was derived from the Natural Resource Damage Assessment Model for Coastal and Marine Environments (NRDAM/CME) (French et al., 1996), which was developed for the U.S. Department of the Interior as the basis of the Comprehensive Environmental Response, Compensation and Liability Act of 1980 (CERCLA) Natural Resource Damage Assessment regulations (as amended) for Type A. These comprehensive oil spill modeling tools have been developed to predict the trajectory (i.e., movement) and fate (i.e., behavior and weathering) of released hydrocarbons into water. The state-of-the-art capability of the oil fate and impact assessment model SIMAP is ideally suited to simulate spill scenarios from PPWs (French et al., 1996; French McCay & Rowe, 2004; French McCay et al., 2004). For wrecks known or suspected to contain chemical hazards, the CHEMMAP model can be applied to simulate spills and measure impacts in a similar fashion to the oil spill modeling but for chemicals which have different behaviors once released into the environment (French McCay, 2001). Using the data from the trajectory, fate, and effects modeling in SIMAP and CHEMMAP, with the addition of the appropriate spill response within the model simulation, the costs and damages from the resulting oil and chemical spills are then calculated. This methodology has been applied in several studies (Etkin et al., 2003, 2006; Etkin & Welch, 2005; French McCay et al., 2005). In addition to the RULET study, oil spill modeling tools have been used to assess the risk from PPW, specifically, and helped to prioritise oil removal operations (Etkin et al., 2009; Etkin, 2019).

Oil spill modeling tools incorporate site-specific and season-specific geographic and environmental variables (e.g., winds, currents, tides, waves, habitat data, bathymetry, temperature, salinity, etc.), as well as the product-specific chemical and physical characteristics of the hydrocarbons to effectively predict its movement and behavior in the environment (Fig. 3.1). They use wind data, current data, and transport and weathering algorithms to calculate the mass of oil components in various environmental compartments (water surface, shoreline, water column, atmosphere, sediments, etc.), oil trajectory over time, surface oil distribution, and concentrations of the oil components in water and sediments. SIMAP also contains physical fate and biological effects models, which estimate exposure and impact on each habitat and wildlife species (or species group) in the area of the spill. Environmental, geographical, physical-chemical, and biological databases supply required information to the model for computation of fates and effects. Seasonal and spatial variations in wildlife populations, particularly migratory birds (which may or may not be present at specific points throughout the year), are incorporated into the modeling databases when available. SIMAP and CHEMMAP can therefore be used to determine the potential biological effects (i.e., acute mortality) that may result following a release of oil and/or chemicals.

3.6   Acute Biological Effects

Biological effects of oil spills can be assessed using the predicted trajectory and fate of hydrocarbon contamination to use the spatially and time-varying concentrations and duration of exposure to determine acute mortality following a release. In the SIMAP model, aquatic biota (e.g., fish, invertebrates) are affected by dissolved hydrocarbon concentrations in the water or sediment. This rationale is supported by the fact that soluble aromatics are the most toxic constituents of oil

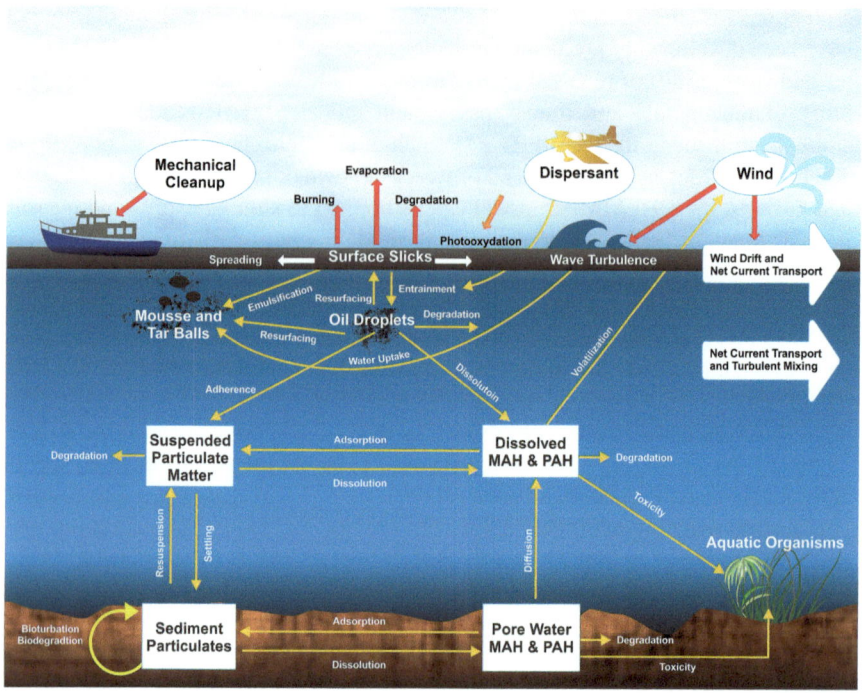

Fig. 3.1 Diagram depicting the transport, fate, and select response processes that may be relevant for subsurface releases from PPW that are simulated in OILMAP and SIMAP

(Neff et al., 1976; Rice et al., 1977; Tatem et al., 1978; Neff & Anderson, 1981; Malins & Hodgins, 1981; National Research Council (NRC), 1985, 2003; Anderson, 1985; French-McCay, 2002). Soluble aromatics refer to low molecular weight compounds that are composed of aromatic rings (six-carbon rings) and can dissolve in water, including compounds such as benzene, toluene, ethylbenzene, and xylene (BTEX), mono-cyclic aromatic hydrocarbons (MAHs) that have 8–10 carbon atoms, and to some extent some smaller poly-cyclic aromatic hydrocarbons (PAHs) that have 10–12 carbon atoms. Exposures in the water column are short, and effects are the result of acute toxicity. In the sediments, exposure can be both acute and chronic, as the concentrations may remain elevated for longer periods of time.

French-McCay (2002) provides estimates of $LC50_\infty$ (which is the infinite time exposure) for MAH and PAH mixtures in fuel and crude oils for spills under different environmental conditions. Figure 3.2 plots LC50 values for total dissolved PAHs for species of average sensitivity under turbulent conditions ($LC50_\infty = 50$ μg/L) for a range of exposure durations and temperatures. The $LC50_\infty$ for 95% of species fall in the range 5–400 μg/L. This oil toxicity model has been validated using laboratory oil bioassay data (French-McCay, 2002). In SIMAP, $LC50_\infty$ for the dissolved hydrocarbon mixture of the spilled oil is input to the model. For each aquatic biota

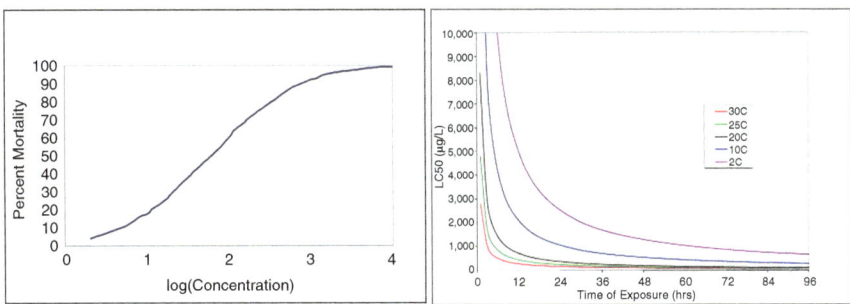

Fig. 3.2 Illustration of percent mortality as a function of concentration (left). The LC50 is at the center of the log-normal function, aligning with 50% mortality. The LC50 of dissolved oil PAH mixtures as a function of exposure duration and temperature (right)

behavior group, the model evaluates exposure duration and corrects the LC50 for time of exposure and temperature to calculate mortality (Fig. 3.2). The oil toxicity model is described in the next section, and in detail in French-McCay (2002). However, it is important to note that an $LC50_\infty = 50$ μg/L does *not* correlate to 50% mortality at a concentration of 50 μg/L if the exposure is for less than several days (unlikely in real world offshore releases). Narcosis from hydrocarbons is generally disruptive to cellular function, but sublethal effects range widely including impairment, decreased growth rate and reproduction, developmental delays, observable structural (deformation), genetic, biochemical, physiological, and behavioral changes. The SIMAP model accounts for the duration of exposure and scales the predicted mortality to lower values for shorter durations of exposure (e.g., hours), which would be more typical in open ocean environments.

The SIMAP biological exposure model estimates the volume and area of water affected by surface oil, concentrations of oil components in the water, and sediment contamination. The exposure model takes into account the time and temperature of exposure. Time of exposure is evaluated by tracking movements of organisms' relative concentrations greater than the concentration lethal to 1% of exposed organisms (LC1, approximated as 1% of $LC50_\infty$, which is the infinite time exposure). Stationary or moving Lagrangian tracers that represent organisms record the concentrations of exposure over time and the dose (summed concentration times duration) to an organism represented by that behavior. Then, the SIMAP biological effects model estimates losses resulting from acute exposure after a spill (i.e., losses at the time of the spill and while acutely toxic concentrations remain in the environment) in terms of direct mortality. Exposure time is the total time concentration exceeds LC1. The concentration is the average over that time, or total dose divided by exposure time. The percentage mortality is then calculated using the log-normal function centered on the LC50 at any specific point in time. The end result is a predicted percent mortality at each point in space and for each representative type of organism and behavior group.

3.7   Seasonal Considerations

As noted above, there are many variables that may impact the magnitude of potential consequences following a release, including: the number of wrecks, their locations, the fuel types and volumes contained, as well as the largest unknown being the specific point in time that a release takes place. One clear example of the point in time of a release impacting the potential for effects is for wrecks within the Great Lakes. Should a release occur in August, the oil would most likely rise to the surface and be transported by winds and currents to adjacent shorelines, with a portion of the oil evaporating to the atmosphere and a smaller portion dissolving into the water column. However, if that same release were to occur in February, the oil may become trapped beneath the layer of ice at the water surface, which would enable much larger portions of the oil to dissolve into the water column as the evaporation of lighter ends would be capped by the ice. In addition, the areal extent of effects may be smaller during the winter, as the oil would become trapped beneath the ice and not transported by surface winds. This would decrease the overall footprint or extent of effects but may increase the magnitude of contaminants and duration of exposure to local organisms, thereby increasing the potential for effects (e.g., acute mortality) to nearby aquatic species. Similar variability can occur in the offshore marine environment, where metocean (i.e., meteorological and oceanic) conditions can be extremely variable in any given time period, even on hourly timescales or below. Therefore, it is important to have an understanding of not only the conditions that may be present at the time of release, but throughout any given year and over many years to have a more complete understanding of the variability in potential transport and fate of any given release. Stochastic assessments will be discussed below to address this environmental variability and the resulting likelihood that releases may impact certain areas.

3.8   Validation Studies

The SIMAP transport model has been validated with more than 20 case studies, including the Exxon Valdez and other large spills such as *Deepwater Horizon* in the Gulf of Mexico (French-McCay, 2003, 2004; French McCay & Rowe, 2004; French-McCay et al., 2018a, b, c, 2021a, b, c). In addition, the models have been further refined and validated using test spills designed to verify the model's transport algorithms (French & Rines 1997, French-McCay et al. 2007) as well as numerous laboratory studies. The oil toxicity model was validated using laboratory oil bioassay data for both fresh and salt-water organisms (French-McCay, 2002) and for lobster mortality in the case of the North Cape spill (French-McCay, 2003). The underlying toxicity data used to develop the model is based on bioassays for freshwater and marine fish, invertebrate, and algal species at a variety of life stages. Below is a summary of the oil toxicity analysis by French-McCay (2001, 2002).

These computational spill modeling tools have been used extensively throughout the world to meet and exceed regulatory requirements for offshore exploration, development, operations, transport, and risk assessment. These tools, and SIMAP specifically, have been used in numerous Natural Resource Damage Assessments (NRDA) for the U.S. government, including notable releases such as the *Deepwater Horizon* oil spill in the Gulf of Mexico beginning on April 20, 2010 (USDOI, 2023).

The *Deepwater Horizon* was a particularly notable release, not only for its magnitude, but because of the large amount of attention that was focused on the release. While the well was still releasing oil, BP dedicated approximately $500 million to be spent over 10 years 'to fund an independent research program designed to study the impact of the oil spill and its associated response on the environment and public health in the Gulf of Mexico.' This investment spawned the Gulf of Mexico Research Initiative, or GOMRI, which is governed by an independent, academic research board of 20 science, public health, and research administration experts and independent of BP's influence. In addition, hundreds of millions of dollars were spent on the scientific study of the response itself by both academia and industry. One result of this was a tremendous focus on improving the understanding of trajectory, fate, and effects modeling. Numerous improvements were made to underlying model algorithms including entrainment, emulsification, weathering, degradation, and other processes that serve move the oil within the environment or alter its chemical and physical state. This spawned numerous studies and improvements in computational oil spill modeling, including several associated with characterising the mass balance of the release, the state of the oil, and the validation of the trajectory, fate, and ultimate effects (French-McCay et al., 2018a, b, c, 2021a, b, c).

3.9 Stochastic Assessments

One of the largest challenges with assessing the risk (and specifically consequence) of a release from a PPW is that the exact time of the release, and the corresponding environmental conditions at that time, are not known. A stochastic approach can be used to determine the footprint and probability of areas that are at increased risk of oil exposure based upon the variability of meteorological and hydrodynamic conditions that might prevail during and after a release. A stochastic scenario is a statistical analysis of results generated from many (e.g., >100 simulations) different individual trajectories of the same release scenario, with each trajectory beginning at a different point in time, selected from a relatively long-term window of time (e.g., 10 years). Stochastic simulations therefore provide insight into the probable behavior of potential oil spills in response to spatially and temporally varying meteorological and oceanographic conditions around a release location (Fig. 3.3). This stochastic approach therefore allows for the same type of release to be analyzed under varying environmental conditions (e.g., summer vs. winter, calm vs. windy, or low vs. high current from 1 year to the next). The results provide the probable behavior of the potential releases based upon this environmental variability.

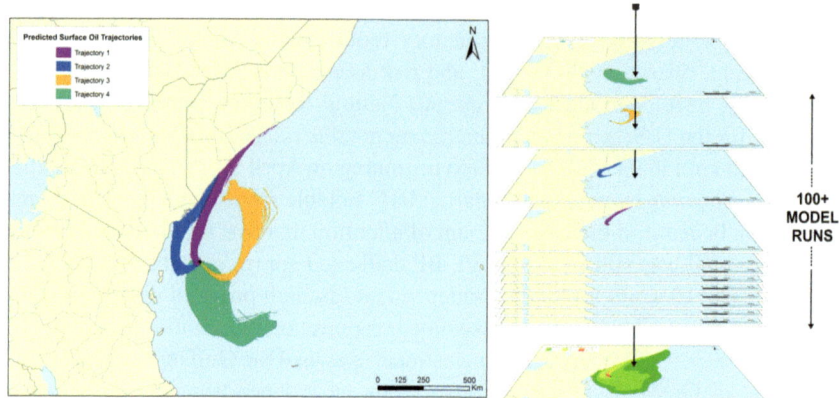

Examples of four individual spill trajectories predicted by OILMAP for a particular spill scenario. The frequency of contact with given locations is used to calculate the probability of impacts during a spill. Essentially, all 100+ model runs are overlain (shown as the stacked runs on the right) and the number of times that trajectory reaches a given location is used to calculate the probability in that location.

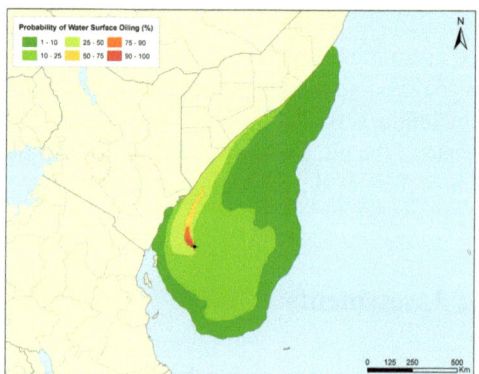

Fig. 3.3 Stochastic modeling approach used to generate probability of surface oil exceeding a given threshold for an example release. The example in the top left highlights four individual release trajectories predicted by SIMAP for a single generic release scenario at a generic location simulated with different start dates and therefore environmental conditions. In a stochastic analysis, over 100 individual trajectories are overlaid (shown as the stacked simulations on the right) and the frequency of threshold exceedance at each location is used to calculate the predicted probability following a release (bottom)

The stochastic model computes surface and subsurface trajectories for an ensemble of hundreds of individual cases for each release, thus sampling the variability in regional and seasonal wind and current forcings by starting the simulation at different dates and times within the timeframe of interest. The stochastic analysis provides two types of information: (1) the footprint of areas that might be oiled and the associated probability of oiling, and (2) the shortest time required for oil to reach any point within the areas predicted to be oiled. The areas, probabilities, and minimum timing of oiling are generated by a statistical analysis of all the individual stochastic runs. The left panel of Fig. 3.3 depicts four individual trajectories predicted by SIMAP for a generic example scenario. Because these trajectories were started on different dates and times, they experienced varying environmental conditions, and thus traveled in different directions. To compute the stochastic results, hundreds of individual trajectories like the four depicted here were overlaid upon one another and the number of times that each given location throughout the modeled domain was intersected by the different trajectories was used to calculate the probability of oil exposure for each specific location. This process is illustrated by the stacked runs in the right panel of Fig. 3.3. The predicted footprint is the cumulative oil-exposed area for all of hundreds of individual releases combined. The colour-coding represents a statistical analysis of all the individual trajectories to predict the probability of oil at each point in space, based upon the environmental variability.

It is important to note that although large footprints of oil would be depicted by a stochastic analysis, they are not the expected distribution of oil from any single discharge. These maps do not provide any information on the quantity of oil in a given area. They simply denote the probability or minimum time of oil exceeding the specific threshold passing through each grid cell location in the model domain at any point over the entire model duration, based on the entire ensemble of simulations. The footprint of any single release of oil, be it modelled or real, would be much smaller than the cumulative footprint of all the runs used in the stochastic analysis. Similarly, the footprint of oil from any individual release at a single time step (snapshot in time) would be even smaller than the cumulative swept area depicted here.

3.10 Conclusions

While the use of computational oil spill modeling tools in the realm of PPW is not new, there have been numerous improvements in recent years that refine our understanding of not only the likelihood of releases, but also the potential movement, behavior, and resulting effects that may follow any specific release. These refinements do affect the predicted trajectory, fate, and effects following a release. When conducting risk assessments, it is necessary to understand both the likelihood of a release as well as the resulting consequence. Computational oil spill modeling can be used to quantify the range of potential consequences, while also informing potential response strategies. Specifically, oil spill models may be used to identify locations where oil may be transported and the range of time for oil to reach those locations, based upon site-specific and season-specific conditions. This will identify key locations of concern (e.g., populated areas, fisheries, sensitive receptors, etc.) that have the potential to be impacted, as well as the associated time of first arrival and duration of exposure to contaminants. Continued focus on modeling of PPW and specific wrecks of interest can help prioritise which wrecks to remediate first

(Brennan et al., 2023). In addition to forecast predictions of any potential future release, the oil spill modeling tools themselves can and have been used in 'reverse' to provide hindcast predictions of the origin location of any observed slicks that may be on the water surface (French McCay, 2001). In essence, winds and currents are reversed to ascertain where a slick may have come from. Therefore, following any observation of oil, oil spill modeling tools can be used to determine whether a known wreck had the potential to be the source of a release. The application of oil spill modeling tools is great in determining the range of potential risk and location of oil contamination, as well as range of potential consequence following a release.

References

Alcaro, L., Amato, E., Cabioch, F., Farchi, C., & Gouriou, V. (2007). DEEPP Project: Development of European Guidelines for Potentially Polluting Shipwrecks, ICRAM, Instituto Centrale per la Ricerca scientifica e tecnologica Applicata al Mare, CEDRE, Centre de Documentation de Recherché et d'Epérimentations sur les pollutions accidentelles des eaux. D.G. Environment, Civil Protection Unit, Contract No. 07.030900/2004/395842/SUB/A5, September 2007, 163 p.

Anderson, J. W. (1985). Toxicity of dispersed and undispersed Prudhoe Bay crude oil fractions to shrimp, fish, and their larvae. American Petroleum Institute Publication No. 4441, Washington, DC., USA, August 1985, 52p.

Brennan, M. L., Delgado, J. P., Jozsef, A., Marx, D. E., & Bierwagen, M. (2023). Site formation processes and pollution risk mitigation of World War II oil tanker shipwrecks: *Coimbra* and *Munger T. Ball. Journal of Maritime Archaeology, 18*, 321–335.

Etkin, D. S. (2019). Developments in risk assessments for potentially-polluting sunken vessels. In *Proceedings of the forty-second Arctic and Marine Oil Spill Program (AMOP) Technical Seminar, Environment Canada, Ottawa, ON*, pp 171–184.

Etkin, D. S., & Welch, J. (2005). Development of an oil spill response cost - Effectiveness analytical tool. In *Proceedings of the twenty-eighth arctic and marine oilspill program technical seminar* (pp. 889–922).

Etkin, D. S., McCay, D. F., & Rowe, J. (2006). Modeling to evaluate effectiveness of variations in spill response strategy. In *Proceedings of the twenty-ninth arctic and marine oil spill program technical seminar* (pp. 879–892).

Etkin, D.S., D. French McCay, J. Jennings, N. Whittier, S. Subbayya, W. Saunders, & Dalton, C. (2003). Financial implications of hypothetical San Francisco bay oil spill scenarios: Response, socio-economic, and natural resource damage costs. *Proceedings of 2003 international oil spill conference, 1*,317–1,325.

Etkin, D. S., Van Rooij, J. A. C. H., & French McCay, D. (2009). Risk assessment modeling approach for the prioritization of oil removal operations from sunken wrecks. In *Proceedings of the thirty-second Arctic and Marine Oil Spill Program (AMOP) Technical Seminar, Environment Canada, Ottawa, ON.*

Etkin, D., French McKay, D., Horn, M., Wolford, A., Landquist, H., & Hassellöv, I. (2017). Chapter 2: Quantification of Oil Spill Risk. In *Oil spill science and technology* (2nd ed., pp. 71–183). Elsevier Publishing. Merv Fingas ed. https://doi.org/10.1016/B978-0-12-809413-6.00002-3

French McCay, D. P. (2001). Chemical Spill Model (CHEMMAP) for forecasts/hindcasts and environmental risk assessment. In *Proceedings of the twenty-fourth Arctic and Marine Oil Spill Program (AMOP) Technical Seminar, Environment Canada, Ottawa, ON*, pp. 825–846.

French McCay, D. P. (2002). Development and application of an oil toxicity and exposure model, OilToxEx. *Environmental Toxicology and Chemistry, 21*(10), 2080–2094. https://doi.org/10.1002/etc.5620211011

French McCay, D. P. (2003). Development and application of damage assessment modeling: Example assessment for the *North Cape* oil spill. *Marine Pollution Bulletin, 47*(9–12), 341–359.

French McCay, D. P. (2004). Oil spill impact modeling: Development and validation. *Environmental Toxicology and Chemistry, 23*(10), 2441–2456.

French McCay, D.P., Rowe, J.J. (2004). Evaluation of bird impacts in historical oil spill cases using the SIMAP oil spill model. In *Proceedings of the twenty-seventh Arctic and Marine Oil Spill Program (AMOP) Technical Seminar, Environment Canada, Ottawa, ON*, pp. 421–452.

French McCay, D., Reich, D., Michel, J., Etkin, D., Symons, L., Helton, D., & Wagner, J. (2012). Oil spill consequence analyses of potentially-polluting shipwrecks. Paper in *Proceedings of the 35th AMOP Technical Seminar on Environmental Contamination and Response, Emergencies Science Division, Environment Canada, Ottawa, ON, Canada*.

French-McCay, D., Reich, D., Michel, J., Etkin, D. S., Symons, L., Helton, D., & Wagner, J. (2014). For response planning: Predictive environmental contamination resulting from oil leakage from sunken vessels. Poster in *Proceedings of the international oil spill conference*, May 2014.

French-McCay, D., Jayko, K., Li, Z., Horn, M., Isaji, T., & Spaulding, M. (2018). Volume II: Appendix II – Oil transport and fates model technical manual. In Galagan, C. W., French-McCay, D., Rowe, J., & McStay, L. (Eds.), *Simulation modeling of ocean circulation and oil spills in the Gulf of Mexico*. Prepared by RPS ASA for the US Department of the Interior, Bureau of Ocean Energy Management, Gulf of Mexico OCS Region, New Orleans, LA. OCS Study BOEM 2018-040; 422 p. OCS Study BOEM 2018-040. Obligation No.: M11PC00028. https://espis.boem.gov/final reports/BOEM_2018- 040.pdf

French, D. P., & Rines, H. (1997). Validation and use of spill impact modeling for impact assessment. In *Proceedings of the 1997 International Oil Spill Conference* (pp. 829–834). American Petroleum Institute.

French, D., Reed, M., Jayko, K., Feng, S., Rines, H., Pavignano, S., Isaji, T., Puckett, S., Keller, A., French III, F. W., Gifford, D., McCue, J., Brown, G., MacDonald, E., Quirk, J., Natzke, S., Bishop, R., Welsh, M., Phillips, M., & Ingram, B. S. (1996). Final report. The CERCLA type A Natural Resource Damage Assessment Model for Coastal and Marine Environments (NRDAM/CME), Technical Documentation, Vol. I – V. Office of Environmental Policy and Compliance, U.S. Department of the Interior, Washington, DC, Contract No. 14-0001-91-C-11.

French-McCay, D., N. Whittier, S. Sankaranarayanan, J. Jennings, & Etkin, D. S. (2004). Estimation of potential impacts and natural resource damages of oil. *Journal of Hazardous Materials, 107*(1–2), 11–25.

French-McCay, D. P., Rowe, J. J., Whittier, N., Sankaranarayanan, S., Etkin, D. S., & Pilkey-Jarvis, L. (2005). Evaluation of the consequences of various response options using modeling of fate, effects, and NRDA costs of oil spills into Washington waters. In *Proceedings of the international oil spill conference*, May 2005, Paper 395.

French-McCay, D.P., Mueller, C., Jayko, K., Longval, B., Schroeder, M., Payne, J. R., Terrill, E., Carter, M., Otero, M., Kim, S. Y., Nordhausen, W., Lampinen, M., & Ohlmann, C. (2007). Evaluation of field-collected data measuring fluorescein dye movements and dispersion for dispersed oil transport modeling. In *Proceedings of the 30th Arctic and Marine Oil Spill Program (AMOP) Technical Seminar, Environment Canada, Ottawa, ON*, pp. 713–775.

French-McCay, D. P., Horn, M., Li, Z., Jayko, K., Spaulding, M., Crowley, D., & Mendelsohn, D. (2018a). Modeling distribution, fate, and concentrations of deepwater horizon oil in subsurface waters of the Gulf of Mexico. Chapter 31. In S. A. Stout & Z. Wang (Eds.), *Oil spill environmental forensics case studies* (pp. 683–736). Elsevier. ISBN: 978-O-12-804434-6.

French-McCay, D.P., Horn, M., Li, Z., Crowley, D., Spaulding, M., Mendelsohn, D., Jayko, K., Kim, Y., Isaji, T., Fontenault, J., Shmookler, R., & Rowe, J. (2018b). Simulation modeling of ocean circulation and oil spills in the Gulf of Mexico. Volume III: Data collection, analysis and model validation. US Department of the Interior, Bureau of Ocean Energy Management, Gulf of Mexico OCS Region, New Orleans, LA. OCS Study BOEM 2018-041; 313 p. Obligation No.: M11PC00028. https://espis.boem.gov/final reports/BOEM_2018-041.pdf

French-McCay, D. P., Jayko, K., Li, Z., Spaulding, M., Crowley, D., Mendelsohn, D., Horn, M., Isaji, T., Kim, Y. H., Fontenault, J., & Rowe, J. (2021a). Oil fate and mass balance for the Deepwater Horizon oil spill. *Marine Pollution Bulletin, 171*, 112681.

French-McCay, D. P., Robinson, H. J., Spaulding, M. L., Li, Z., Horn, M., Gloekler, M. D., Kim, Y. H., Crowley, D., & Mendelsohn, D. (2021b). Validation of oil fate and mass balance for the Deepwater Horizon oil spill: Evaluation of water column partitioning. *Marine Pollution Bulletin, 173*, 113064.

French-McCay, D. P., Spaulding, M. L., Crowley, D., Mendelsohn, D., Fontenault, J., & Horn, M. (2021c). Validation of Oil Trajectory and Fate Modeling of the Deepwater Horizon Oil Spill. *Frontiers in Marine Science*. 37 pp. FMARS. 2021.618463.

Landquist, H., Rosén, L., Lindhe, A., Norberg, T., Hassellöv, I.-M., Lindgren, J. F., & Dahllöf, I. (2014). A fault tree model to assess probability of contaminant discharge from shipwrecks. *Marine Pollution Bulletin, 88*(1–2), 239–248.

Malins, D. C., & Hodgins, H. O. (1981). Petroleum and marine fishes: A review of uptake, disposition, and effects. *Environmental Science & Technology, 15*(11), 1272–1280.

Michel, J., Etkin, D., Gilbert, T., Urban, R., Waldron, J., & Blocksidge, C. (2005). Potentially polluting wrecks in marine waters. *Paper in Proceedings of the 2005 international oil spill conference*, May 2005.

National Oceanic and Atmospheric Administration. (2013). *Risk assessment for potentially polluting wrecks in U.S. Waters*. National Oceanic and Atmospheric Administration, Silver Spring, MD. 127 pp. + appendices. Prepared by U.S. Department of Commerce, National Oceanic and Atmospheric Administration, Office of National Marine Sanctuaries, Office of Response and Restoration, RPI, ASA, and Environmental Research Consulting. March 2013.

National Research Council (NRC). (1985). *Oil in the sea: Inputs, fates and effects* (601p). National Academy Press.

National Research Council (NRC). (2003). *Oil in the sea III: Inputs, fates and effects*. National Academy Press. 280p.

Neff, J. M., & Anderson, J. W. (1981). *Response of marine animals to petroleum and specific petroleum hydrocarbons* (177p). Applied Science Publishers Ltd./Halsted Press Division, Wiley.

Neff, J. M., Anderson, J. W., Cox, B. A., Laughlin, R. B., Jr., Rossi, S. S., & Tatem, H. E. (1976). Effects of petroleum on survival respiration, and growth of marine animals. In *Sources, effects and sinks of hydrocarbons in the aquatic environment* (pp. 515–539). American Institute of Biological Sciences.

Overfield, M. L. (2004). Resources and UnderSea Threats (RUST) Database: An assessment tool for identifying and evaluation submerged hazards within the national marine sanctuaries. *Marine Technology Society Journal, 38*(3), 72–77.

Rice, S. D., Short, J. W., & Karinen, J. F. (1977). Comparative oil toxicity and comparative animal sensitivity. In D. A. Wolfe (Ed.), *Fate and effects of petroleum hydrocarbons in marine ecosystems and organisms* (pp. 78–94). Pergamon Press.

Symons, L., Michel, J., Delgado, J., Reich, D., French McCay, D., Etkin, D., & Helton, D. (2014). The Remediation of Underwater Legacy Environmental Threats (RULET) Risk assessment for potentially polluting shipwrecks in U.S. Waters. *Paper in Proceedings of the 2014 international oil spill conference*, May 2014.

Tatem, H. E., Cox, B. A., & Anderson, J. W. (1978). The toxicity of oils and petroleum hydrocarbons to estuarine crustaceans. *Estuarine and Coastal Marine Science, 6*, 365–373.

United States Department of the Interior. Accessed 15 July 2023. Deepwater Horizon Response and Restoration: Administrative Record. Retrieved from https://www.doi.gov/deepwaterhorizon/adminrecord

Zelo, I., Overfield, M., & Helton, D. (2005). NOAA's abandoned vessel program and resources and under sea threats project—Partnerships and progress for abandoned vessel management. *NOAA Restoration and Response. International Oil Spill Conference Proceedings, 2005*(1), 807–808.

Open Access   This chapter is licensed under the terms of the Creative Commons Attribution 4.0 International License (http://creativecommons.org/licenses/by/4.0/), which permits use, sharing, adaptation, distribution and reproduction in any medium or format, as long as you give appropriate credit to the original author(s) and the source, provide a link to the Creative Commons license and indicate if changes were made.

The images or other third party material in this chapter are included in the chapter's Creative Commons license, unless indicated otherwise in a credit line to the material. If material is not included in the chapter's Creative Commons license and your intended use is not permitted by statutory regulation or exceeds the permitted use, you will need to obtain permission directly from the copyright holder.

Chapter 4
Corrosion Processes of Steel-Hulled Potentially Polluting Wrecks

Robert Glover

4.1 Overview

Submerged metals are continuously affected by the chemical processes of corrosion, the destructive degradation of metal by chemical or electrochemical reactions within the marine environment (Valenca et al., 2022:2–3; Venugopal, 1994:35). Over time, metal ions at anodic sites defuse into electrolytic solutions from the oxidising reactions occurring at cathodic sites, causing the creation of corrosion byproducts, like rust on iron, and the loss of structural mass. The different reduction reactions in the microstructures of alloys and the imperfections found within refined materials, like carbon slag in steel, are targeted by this process, essentially reverting the chemically unstable materials back to their more stable original forms (Moore III, 2015:192; MacLeod, 2016a:90–92). The deterioration of metallic shipwreck hulls has become a growing concern within the field of marine conservation as many of the fuel tankers deliberately targeted in WWII threaten to release trapped fuel and chemical cargoes after nearly eight decades of exposure to a range of corrosive environments (Barrett, 2011:4–5). With the deterioration rate of ship hulls averaging at around ±0.1–0.4 mm of loss per year and the thickness of ship deck plates from the 1940s to the 1960s generally ranging from 1–4 cm in thickness, the window to act on the majority of potentially polluting shipwrecks (PPW) before a catastrophic breach occurs is closing (MacLeod, 2016b:8; Beldowski, 2018:249; Masetti, 2012:33; Masetti & Calder, 2014:139).

R. Glover (✉)
University of Southampton, Southampton, UK
e-mail: rsg1e22@soton.ac.uk

© The Author(s) 2024
M. L. Brennan (ed.), *Threats to Our Ocean Heritage: Potentially Polluting Wrecks*, SpringerBriefs in Underwater Archaeology,
https://doi.org/10.1007/978-3-031-57960-8_4

While laboratory and field methods of corrosion analysis have become well-established and can provide researchers with malleable models for understanding steel corrosion, a definitive predictive system for the manner of and time to failure of shipwreck hulls remains challenging (Russell et al., 2004:37). Wreck sites are open systems, with constant exchanges of materials and energy creating a state of dynamic negative disequilibrium unique to every environment and situation, but which will all ultimately result in the complete disintegration of the hull (Quinn, 2006:1420). While the full scope of factors affecting a shipwreck is too large to definitively quantify, a wide understanding of the host of corrosion factors encountered *in situ*, as well as an understanding of the increasing complexity created by structural degradation, is useful when attempting to decide a course of action for a specific wreck (Etkin et al., 2009:3–4; Russell et al., 2004:37). Inevitably, the metal hull of every PPW will corrode to a point of partial or complete failure, allowing oil or other pollutants to escape into the local environment. It is therefore advantageous to have as thorough an understanding as possible of the factors unique to each site and vessel that will accelerate or reduce corrosion rates. This will allow for the development of a timeline to failure for each ship, which is essential to remediating the wreck before it creates an environmental catastrophe.

4.2   Corrosion Types in Shipwrecks

Corrosion of shipwreck metals can be very loosely grouped into two categories: relatively uniform corrosion found across an exposed surface and more harmful, localised corrosive attacks on specific portions of the hull (Tait, 2012:864–865; Yongjun Tan, 2023:1). Uniform corrosion, otherwise known as general corrosion, is typically characterised by the gradual thinning of surfaces and protective corrosion products by the abrasive electrochemical and mechanical characteristics or contents of a surrounding electrolytic solution (Xia et al., 2021:2). A short period of rapid corrosion due to high rates of oxygen diffusion along the exposed surface will occur upon initial submersion which will then be followed by a slower, longer-term corrosion rate protected by the buildup of corrosion products, biofilms, sand particles, and static organisms, commonly called concretions (Melchers, 2003:272). Concretions form semi-permeable, anaerobic barriers between the bare metal and the seawater, creating a protective outer coating from oxygen diffusion corrosion while also containing heightened acidity and chloride ion concentrations, stabilising the hull at a steady deterioration rate (North, 1976:254–257). Uniform corrosion will not affect the structural integrity of the metal until a large portion of the cross-section has deteriorated, making it a reliable and stable way to measure, predict decay rates, and create corrosive models for steel-hulled vessels (Nürnberger et al., 2007:195; MacLeod & Viduka, 2011:135). Corrosion rates can drastically increase, however, if the concretion layer protecting the hull is stripped off or broken in any way, as the metal becomes exposed again to the marine environment and vulnerable to rapid, targeted attacks until recolonised (MacLeod et al., 2017:270–280).

Microbiologically influenced corrosion (MIC) is a form of corrosion that can develop underneath, or even help to build, a concretion. It can be difficult to predict the effect that a biofilm, a conglomeration of synergistic communities of microorganisms, will have on the surface that it is found on. On the one hand, biofilms can reduce corrosive potential by decreasing oxygen concentrations, halting diffusion by active transport, and producing corrosion inhibitors. On the other hand, however, biofilm communities have also been observed producing oxygen, sulfides, ammonia, and highly concentrated acids, using various structural metals as electron donors in metabolic processes, ennobling metals to produce galvanic couples, altering anions to a more aggressive, corrosive state, deactivating corrosion inhibitors, and deriving energy by oxidising metals, all processes that accelerate the corrosion of metals (Little & Lee, 2009:2–21). Ultimately, whether a microbial community is going to protect or target a hull is going to depend on the type of metal that the MIC is found on, the biofilm community itself, and the surrounding electrolytic solution in which the vessel is submerged, with some seasonal variation seen in certain locations (Little & Lee, 2009:2–3).

Localised corrosion can be initiated in the natural environment in many ways, as described by Galvele (1983:2). Corrosion from the metal's electrochemical reaction with the environment is thought to be initiated by the chemical breakdown or mechanical disruption of the protective oxide film of a metal or alloy. Aggressive anions can become trapped in this small break between the film and the surface of the metal, with corrosion rates reaching several thousand to millions of times higher than in the surrounding passive metal, causing serious sectional and penetrative damage to the material structure (Kruger & Rhyne, 1982:206–207; Vargel, 2020:164–166; Melchers, 1999:6). Pitting corrosion, one of the most common and harmful types of localised chemical attack, is confined to these small breaks, on the order of square millimeters or less, and is covered with corrosion products that restrict ionic species flow in or out of the pit, drastically increasing the corrosion rate (Kruger & Rhyne, 1982:206–207; Galvele, 1983:1–15). Crevice corrosion is a similar chemical attack that occurs under corrosion product deposits and is found in component crevices, like structural support couplings or joints (Kruger & Rhyne, 1982:206–207; Makhlouf, 2015:541–543). Cavitation corrosion is a physical corrosive attack initiated by the environment and is caused by the collapse of gas bubbles on the surface of the metal. The subsequent increased velocity of the electrolyte fluid creates miniature shock waves on the surface, thus creating breaks in the protective oxide film and initiating pitting corrosion (Makhlouf et al., 2018:111).

Several forms of localised corrosion occur within the metal itself and can be caused by dynamic pressures or material differences in the components or alloys. Galvanic coupling, also referred to as bimetallic or proximity corrosion, is a static reaction that occurs when metals with different corrosive potentials touch each other or are connected by the same concretion in an electrolytic solution (MacLeod, 2019:871; North, 1984:133–134). While the metal with the higher electro-reactivity potential (*E corr.*) in the coupling will begin to erode more quickly, the lower potential metal will be protected and begin to corrode more slowly (North, 1984:133–134). This reaction between materials has been used as a form of temporary protection for fragile artifacts with the use of sacrificial anodes, as demonstrated by corrosion scientist Ian MacLeod (2016b:9), in collaboration with North (McCarthy, 2000:86–88) and Steyne (MacLeod & Steyne, 2011:347–349; Steyne & MacLeod, 2011:67). Intergranular corrosion is another static, localised attack at and adjacent to grain boundaries between the microstructures of a metal alloy, caused by chemical differences in the grain, impurities at the grain boundaries, and reduction or enrichment of an alloying element in the grain boundary area. This form of corrosion is only prevalent in alloys and can severely impact the material's structural strength or even cause it to disintegrate (Karlsdottir, 2022:259–260).

Stress corrosion cracking, a dynamic attack, can be particularly damaging to a crumbling shipwreck that is experiencing changing environmental pressures. It is characterised by internal, residual stresses, created when the metal is formed, welded, or processed, and external, environmental stresses, applied to a metal component by the environment. This form of corrosion can result in intergranular or transgranular cracking of the source material (Makhlouf et al., 2018:114). Corrosion fatigue, a similar dynamic attack, is caused by repeated, cyclical exposures to a corrosive environment and external stresses. Eventual catastrophic structural collapse can potentially occur at any time, even when those stresses are not presently in action (Komai, 2003:345).

4.3   OCP and General Corrosion Factors

The kinetic driving force behind shipwreck corrosion can be described as the Open Circuit Potential (OCP), or corrosion potential. This value compares the equilibrium potential between a metallic wreck, referred to as the working electrode, and its electrolyte or environment to a reference electrode in the same environment, providing a measurable rate of real corrosion behavior (Siddaiah et al., 2021:7–9; Rasol et al., 2015:294). An increase in OCP results in a depolarisation of the cathode and an increase in corrosion, whereas a decrease in OCP will exhibit the opposite effect (Rasol et al., 2015:294). Well-documented are the physical, chemical, and biological corrosion factors that positively increase corrosion potential and threaten the structural stability of shipwreck hulls (Kuroda et al., 2008:3–6; MacLeod & Viduka, 2011:136; Liddell & Skelhorn, 2019:83; Gilbert et al., 2003:178–179). For example, the type of metal used in the reaction, a mild steel in the case of most PPWs, and its associated microscopic structure, along with the microorganisms forming biofilms on the surface of the material, can influence OCP (Eyres & Bruce, 2012:45–49; Rasol et al., 2015:294–297). Differing areas of the ship may also have unique open circuit potentials due to dissimilar microstructures in the materials or passive film layers, contributing to non-uniform corrosion across the vessel

(Mischler & Munoz, 2018:508). Areas that have been exposed to physical stresses, including physical damage or heat, may change enough in material structure to positively shift the OCP of that component (Moreto et al., 2018:2–5). Hull-rivet interactions, welded seams, or damage due to impact or fire may result in accelerated corrosion of an area or interface relative to an intact, undamaged section of the hull, resulting in a higher risk of corrosion (Mischler & Munoz, 2018:508; Rasol et al., 2015:294–297; Chaves et al., 2022:195). Every shipwreck, however, is being acted upon by a unique set of environmental characteristics that should be properly understood before attempting to create a roadmap to failure for the ship.

4.4 Acute Environmental Factors

In the face of destructive natural events, such as catastrophic storms and sudden coastal changes, metal shipwrecks can suffer damages complicated by their age and preexisting structural degradation. Volcanic tremors affecting shipwrecks, while rarely studied, have been shown to affect site structural integrity by shifting vessels deposited on sloped bedforms. This can produce wear through mechanical damage and abrasion or by exposing shallow water wrecks to corrosion factors related to coastal erosion (North & MacLeod, 1987:74–75; Ridwan, 2019:1624). Events such as earthquakes and subsequent tsunamis can have severe impacts on site dynamics as well, drastically altering the surrounding landscape and exposing hulks not just to more dynamic surface currents and tidal action, but also to more anthropogenic activity in the form of plundering, fishing, vandalism, or accumulated rubbish and waste (Ridwan, 2019:1624–1625). Estuarine and riverine wrecks are prone to experience acute damage during floods and droughts as higher rates of corrosion impact abandoned watercraft found at air-water interfaces (North & MacLeod, 1987:75).

Large tropical storms, cyclones, hurricanes, and typhoons can cause the physical degradation of a wreck by dislodging loose pieces of the vessel from the whole, reinforcing the effects of tidal currents to enact scour damage, altering the composition of the water column, subjecting the structure to abnormally powerful wave action, washing away supportive surrounding sediments, damaging or stripping away concretions, or even causing the collapse of the hull (MacLeod & Viduka, 2011:135; North & MacLeod, 1987:75; Steyne & MacLeod, 2011:68; MacLeod et al., 2017:270). Sudden shifts in the surrounding environment, as well as collapses inside of the wreck due to large weather events, have been shown to create structural damage that can expose new sections of the ship to corrosion, altering existing corrosion patterns. This was seen with damage to the wreck of USS *Mississinewa* caused by a large summer storm in 2001, which subsequently initiated a series of oil leaks that grew in frequency and severity until the wreck had to be relieved of its oil cargo several years later (U.S. Navy, 2004:1-1—1-2). Structural collapses can also cause changes in fluid dynamics around the wreck, which will negatively affect areas of the ship that were previously spared these velocities (MacLeod et al., 2017:273).

4.5   Cyclical Environmental Factors

Where they exist, regular fluctuations in water column characteristics must be considered in models of corrosive behavior. Cycles in the biochemical characteristics of a shipwreck site will regularly occur throughout the year, as salinity, dissolved oxygen, and microbial activity in the water column will experience high and low points due to seasonal changes and the minor associated shifts in ambient water temperature, velocity, and microbial content (Li et al., 2019:6056–6060; Olson et al., 2022:1–2; Mestre et al., 2020:1–2; Liao et al., 2022:4–17). More dramatic site-specific cyclical characteristics, like the variable seasonal temperature changes witnessed at most inshore coastal sites, can profoundly influence corrosion potential (Zintzen et al., 2008:330). Strong water velocities, caused by particularly aggressive tidal action and recurrent storm patterns, can accelerate corrosion or destabilise the vessel, creating leaks. The SS *Jacob Luckenbach* wreck oiled birds and beaches along the San Francisco Bay area for decades because seasonal current variations and regular winter storms disturbed the wreckage, releasing fuel into the water column (Moffatt, 2004:65; Duerr et al., 2016:1). Current flow which is not uniform across the structure may result in variable corrosion rates, reducing the value of predictions made using a uniform corrosion model. The *Cerberus*, for example, offshore Melbourne, Australia, has experienced more deterioration on its starboard side from exposure to the open ocean and the associated higher fluxes of oxygen than on its port side, which faces into Port Philip Bay (MacLeod & Steyne, 2011:341).

Metal-hulled ships permanently or cyclically exposed to the atmosphere by wave action in the tidal zone, like the Civil War era *Sub Marine Explorer* found in the Bay of Panama, will not only experience the physical stresses caused by the crashing of the water, but will additionally undergo periodic wetting and drying, increased oxygen availability at the air-water interface, physical destruction of protective concretions and corrosion products, impacts from dissolved or transported debris, larger variations in temperature, expedited cycles of erosion, and concentrations of aggressive salts, which can all contribute to an advanced decay rate (North & MacLeod, 1987:75; Evans et al., 2009:46; Johnson et al., 2010:58–59). On the other hand, cyclical exposure to an anaerobic environment has been observed contributing to accelerated MIC for certain metals. When seasonal erosion patterns expose and re-cover a shipwreck, corrosive bands are formed, like those seen on the copper wires found in the SS *Xantho*, an abandoned steamship found off the coast of Western Australia (MacLeod, 2002:702; McCarthy, 2000:91–92). Similarly, corrosion rates of wrecks in areas of high seasonal runoff or seasonal hypoxic conditions due to the migration of pollutants or high microbial or marine growth activity can also be expected to vary (North & MacLeod, 1987:75).

4.6   Long-Term Environmental Factors

The natural characteristics of the wrecking location can often be a strong determinant in how long a shipwreck will be preserved. Vessels that are deposited in a high-energy environment will experience more rapid movement of currents and tidal actions across the metal surface, causing the hull to be impacted by a greater flux of dissolved oxygen and preventing the formation of protective corrosive films, causing non-uniform corrosive potential to increase (North & MacLeod, 1987:75; MacLeod et al., 2017:280). In contrast, if seawater is still or stagnant for extended periods, as is often encountered in deep bays, atolls with narrow, shallow entrances, or at very deep-water wrecks, corrosion rates will rapidly decrease (North & MacLeod, 1987:75). Alternatively, deep-water wrecks are prone to developing 'rusticles', or icicle-shaped iron oxide accumulations, that can be found throughout the vessel and seemingly mobilise structural mass to different areas of the ship. Famously having been found in the *Titanic*, the corrosion phenomena, which appears to develop more extensively on Atlantic deep-water wrecks than Pacific, has been theorised to be the result of MIC-related biodeterioration, dissolved iron accumulation deposited on the wreck, or natural corrosion processes interacting with intense hydrostatic pressures (Salazar & Little, 2017:26–30; Cullimore et al., 2002:117–120; Silva-Bedoya et al., 2021:10–23). Seabed composition can also be important for the long-term preservation of a shipwreck. Frequent current and storm action can create bedforms predominantly composed of rocky, gravelly sand and shells, which can be highly abrasive (Wheeler, 2002:1151). Bedforms consisting of larger-sized sand grains and gravel can still be affected by strong bottom currents which can carry sediment to wash the hull, effectively sandblasting it, or covering it and causing it to collapse under the excess weight (Hac, 2018:175; North & MacLeod, 1987:75). Meanwhile, sediments washed away from under a structurally significant portion of the vessel can create additional weights and strains to become apparent on the metal as it is forced to stay rigid, eventually causing that section to buckle and break away (Hac, 2018:175).

Coastlines that are regularly indented, have high bathymetric relief, and numerous estuaries, as seen along Ireland's southern, western, and northern coasts, will provide stable micro-environments that can protect shipwrecks from rough wave action and mechanical corrosion (Wheeler, 2002:1151). Protective atolls and shelters, including Ulithi Atoll, final resting place of USS *Mississinewa*, and Wardang Island, South Australia, the resting place of *Songvaar*, can provide wave-breaks that will preserve the vessel from storm action, current velocity, or strong tidal surge (Moore et al., 2014:21–22; MacLeod, 2002:703). Similar complications can be found on the seabed depending on the topography of the area. A vessel or artifact deposited in a narrow reef gully may experience a higher water velocity and thus a greater corrosion potential, while a wreck on the leeside of a reef may be protected from these same corrosion pressures (MacLeod, 2018:65–68). Alternatively, a crevice or depression may protect a vessel or artifact from high surges and extensive wave action that would normally cause artifacts to physically shift or roll along the seabed, collecting mechanical damage and abrasion (North & MacLeod, 1987:75).

4.7   Anthropogenic Factors

There are a range of anthropogenic activities that can impact the corrosive potential of a metallic shipwreck, with a wide scope of agents responsible for the damages. Individual divers can cause damage by penetrating a wreck and trapping vented oxygen in ship crevices or by grabbing the structure for stability (Ridwan et al., 2014:8–9). Dive companies in the past have been responsible for damaging sites by dropping anchors onto them or by tying off their boats to vessel superstructures while mooring (Viduka, 2011:14; MacLeod & Steyne, 2011:340; Henderson, 2019:8). Dredging and trawling activities can have significant impacts on the seabed and the cultural heritage that is found there (Evans et al., 2009:46). Brennan and colleagues have noted the damages that *Coimbra* suffered from multiple dredge impacts, eventually leading to the formation of a crack in the hull, while Delgado and colleagues discovered multiple sets of trawl gear trapped under the hull of the *Coast Trader* (Brennan et al., 2023:328; Delgado et al., 2018:27). Fishers and anglers who forcefully retrieve gear stuck on a wreck may damage portions of the vessel in their struggle, while abandoned net hangs, called 'ghost nets', can increase dynamic stress on the structure by inducing drag, destabilising more fragile portions of the ship (Firth, 2018:14; Ridwan, 2019:1627). Modern explosion damage is surprisingly common on wrecks in the Pacific, as fishermen will use ordinance to stun and kill fish found around the bountiful artificial reefs while metal scavengers will use explosives to loosen sheet metal for sale (Naughton, 1985:16–17; MacLeod et al., 2017:270; Browne, 2019:2). Larger scale metal scavenging operations have become far more sophisticated, lifting entire wrecks from the seafloor using cranes installed on large ships (Ridwan, 2019:1625). Souvenir hunters can damage a site while pilfering for treasures, robbing the asset of valuable archaeological evidence, while significant amounts of marine litter can ruin the site's aesthetics and potentially cause physical or galvanic damage to artifacts (Viduka, 2011:14; Ridwan, 2019:1625).

Largescale coastal developments and the associated transformations in the local environment can increase risks to submerged heritage assets as well (Ridwan et al., 2014:8–9; Evans et al., 2009:45–46). Some wrecks that pose threats to modern watercraft or shipping, like the *Cleveco,* have been damaged by passing ship traffic, causing oil spills, and have suffered shifting structural pressures while being lifted and moved, causing even more spills (Davin & Witte, 1997:783–786). Coastal expansion projects and offshore energy exploration and infrastructure development have led to the rediscovery of many vessels but could also result in damage to the wrecks themselves, to the surrounding environment, or negative attention from curious divers (Ridwan et al., 2014:8–9; Church et al., 2009:51; Evans et al., 2009:45–46;

Moore, 2021:157). Oil spills that come into contact with a metallic shipwreck can have significant disruptive effects on the diversity of microorganisms and corrosion product communities that are protecting the metal surface as well as on the porosity of the surrounding sediments, increasing corrosive potential (Salerno et al., 2018:4–12; Hamdan et al., 2018:3–12; Zhang et al., 2022:2–9). Trade routes and shipping lanes, while instrumental in the deposition of many cultural heritage sites, are also significant sources of heavy metals, oil, and human waste pollution in the modern era, leading to the further degradation of these historic vessels on the seabed (Lawrence, 2008:6–14; Chan et al., 2001:581–582). Poor solid and liquid waste management from urban and agricultural landscapes and rainy season runoff near these locations can create significantly elevated nutrient concentrations around shipwreck sites, potentially expediting corrosion rates by up to 50% (Jiminez et al., 2017:3; Melchers, 2014:110–115). Finally, as ocean temperatures and acidity levels rise due to the effects of climate change, the associated physical and chemical changes seen in the environment will cause further damage to shipwrecks, as is already being seen with the expedited chemical breakdown and growth of rusticles found on the *Titanic* (Wright, 2016:260–263; Mann, 2012:44–49).

4.8 Ship Construction, Condition, Impact, and Orientation

Hull joinery technique becomes an important factor in the long-term stability of a wreck and its ability to successfully hold a potentially polluting cargo. Hull plate rivets often see differing corrosion potentials between head, shaft, base metal, and hull, causing them to corrode preferentially to the surrounding metals and release seeps of oil or allow hull plates to separate from the body of the ship (Johnson et al., 2011:7; MacLeod & Steyne, 2011:340; Brennan et al., 2023:327). Welded hulls will typically withhold petroleum pollutants more effectively over a longer period but have been observed bursting along the welded seam during particularly violent impacts with the seabed, releasing a cloud of cargo product (Kery & Stauffer, 2015:7). Vessels sunk in shallow waters will have little time in the water column to gather momentum or shift orientation before eventually depositing in the sediment and thus will have a less energetic impact when landing, as compared to deep-water wrecks that will have a more variable deposition orientation and higher velocity upon impact (Kery & Stauffer, 2015:1–2; Liddell & Skelhorn, 2019:83). High-velocity impacts upon the seabed have been shown to cause deformations, including folds and staving, in various portions of the hull, and can burst sealed doors and hatches, exposing more internal spaces to corrosive processes, accelerating dissolved oxygen transport and increasing opportunities for MIC (Morcillo et al., 2004:122–123).

The vessel's working life and sinking event will also significantly affect its structural integrity. Military and tanker vessels will have accrued residual damage throughout their working lifetimes that will influence OCP and are expected to have suffered extensive structural damage during their sinking events (Gilbert et al.,

2003:178–179; McKay, 2005:129). Portions of the vessel that are missing paints or other protective coatings due to blows and scratches, or were intentionally left unpainted, will be targeted first by corrosive processes (Morcillo et al., 2004:125). Vessels that have been broken into multiple pieces will experience a higher percentage chance of inverting or landing on their sides on the seabed, exposing the vessel to a more diverse range of structural pressures (Brennan et al., 2023:325–334; Morcillo et al., 2004:122–123; Russell et al., 2004:36–42). Corrosion cells can also develop when a vessel has been split into multiple pieces, in which portions of the hull will act as a protective anode for other pieces of the ship with lower electroreactivity potentials (Viduka, 2011:16). Unnatural weight distributions or new stresses created by the deterioration of the ship in directions or areas not considered in the original design will cause the corrosion potential of different sections of the ship to increase (Gilbert et al., 2003:178–179; MacLeod, 2016b:8). A vessel that settles upright on the seabed will experience a generally stable rate of decay due to pressures generally in-keeping with design limitations, and, if sunk with a fuel cargo, may slowly release it from fragile vents and risers, eventually allowing seawater to enter the emptied cavities (Brennan et al., Chap. 9, this volume). A vessel that is inverted or on its side will experience strains that it was not designed to handle, depending on both the topography of the seabed and the strength of currents impacting the broadsides and bottoms of the hull, thus experiencing an expedited corrosion rate (MacLeod et al., 2017:280; Kery & Stauffer, 2015:5). Finally, cargo type, viscosity, and amount can play a part in the protection of the fuel tanks withholding a petroleum or chemical cargo. Thicker, more viscous oils, like a bunker fuel, can create a protective coating along the insides of the tank, delaying corrosion, as observed on the *Coimbra* wreck (Brennan et al., 2023:327). Lighter oils and ballast water can create humid, corrosive environments that will be detrimental to the integrity of the tank (Zayed et al., 2018:300).

4.9   A History of Shipwreck Corrosion Analysis

The compilation and breakdown of the previously described site-specific characteristics into data points for use in a single universal shipwreck corrosion model has been a lengthy and challenging endeavor. The first forays into quantifying the complex processes occurring at shipwreck sites began in the 1970s and 1980s with *in-situ* corrosion analysis led by marine archaeologists North and Pearson while they were studying the wreck of *Batavia*. They attempted to collect and analyze artifact concretions through the lens of corrosion science, allowing them to develop theories on iron artifact preservation and graphitisation (North et al., 1976:192–193; North & Pearson, 1978:180–182). McCarthy would build on this momentum in the 1980s with his survey of the *Xantho*, a steamship that sank off the coast of Western Australia, by combining data extracted from the marine environment with data gathered from eyewitness accounts and historical documents (McCarthy, 2000:7–61; Moore III, 2015:194–196). This study was essential to understanding the rate and manner of decay for an iron shipwreck, including non-uniform rates of decay, and the difficulties encountered when attempting to excavate, protect, and recover artifacts (McCarthy, 2000:88–177). The *Xantho* became the first vessel to be protected by sacrificial anodes as well, a method later perfected and used extensively by Australian corrosion scientist Ian MacLeod (McCarthy, 2000:84–88; MacLeod, 1987:50–55, 1989:7–13, 2019:877; Steyne & MacLeod, 2011:67).

MacLeod would become crucial in the interpretation of chemical and physical modes of shipwreck and metallic artifact degradation using on-site measurements. These measurements would ultimately aid him in his development of corrosion rate linear correlation graphs, used to more easily track and display rates of decay, including for specific environmental factors (Moore III, 2015:196–197; MacLeod, 1989: 7–13, 1995:54–58). MacLeod applied many of these techniques to his study of multiple WWII steel vessels found in Chuuk Lagoon, highlighting the importance of depth, vessel position, breakage pattern, and topography when determining corrosion potential (MacLeod et al., 2011:1–10, 2017:270–281). The 1987 assessment of the USS *Monitor*, only possible via ROV due to the submarine's resting depth, produced more data on the electrochemical behavior of a complete shipwreck, examining the galvanic interaction between vessel components as well as the electrical continuity between adjoining sections (Moore III, 2015:197–198; Arnold III et al., 1992:52–55). The multidisciplinary nature of the *Monitor* study, which combined the efforts of the National Oceanic and Atmospheric Administration (NOAA) and the US Navy, would foretell the continuing combination of research interests that would be necessary to gain a more complete understanding of the complex corrosion processes at play on shipwreck hulls. The development of new probabilistic models and equations that investigate specific environmental influences has given scientists and archaeologists many broad, quantitative tools to use, albeit with the false assumption that the shipwreck is corroding at a uniform rate (Moore III, 2015:199–201; Woloszyk & Garbatov, 2022:2–12; Guedes Soares et al., 2011:529–537; Melchers, 2005:2391–2404). These complex, multifaceted webs of study, however, have since been combined to advance understanding of the complex corrosion environment of a single shipwreck: USS *Arizona*.

4.10   Case Study: USS Arizona

The USS *Arizona* is an American battleship that was sunk by the Imperial Japanese Navy on December 7, 1941, in a surprise attack on Pearl Harbor, Oahu, Hawaii (Wilson et al., 2007:14). Due to its historical significance, status as a war grave for over 1170 military personnel, and the approximately 2500 tons of fuel oil trapped within, the vessel has received significant attention in the form of legal protections, surveys, and structural conservation efforts that have allowed it to remain mostly intact since its initial deposition (Johnson et al., 2018:747). With permissions received from the National Park Service and USS *Arizona* Memorial, the ship has become the first long-term research resource for the study of steel hulled ship corrosion in the marine environment (Johnson et al., 2018:747–750; Wilson et al., 2007:14–18; Russell et al., 2004:37–43; Foecke et al., 2010:1091–1100; Russell et al., 2006:311–317; Murphy, 1987:10–15; Johnson et al., 2011:1–8). Since 1999, the USS *Arizona* Preservation Project has brought together experts from a variety of disciplines to understand the complex internal and external corrosion and deterioration processes affecting *Arizona* and how they have impacted the structural integrity of the vessel (Wilson et al., 2007:15; Russell et al., 2004:35). The goal of this interdisciplinary, cumulative, and years-long project has been to model and predict the nature and rate of structural changes up to imminent collapse, to be used by future site managers to minimise environmental hazard from fuel release and inform decisions for long-term preservation (Wilson et al., 2007:15; Russell et al., 2004:35, 2006:310).

Understanding the corrosion rate of the *Arizona* wreck has involved breaking down the major factors affecting shipwreck decomposition into data that can be quantitatively measured, plotted, and modeled (Russell et al., 2004:37). Finite Element Analysis (FEA) modelling has become the principal receptacle into which all experimental and observational corrosion data is fed to create a highly accurate and realistic calculation of the stresses and changes seen in the structure under load (Russell et al., 2004:37). The ship model, divided into sections or 'elements', is mapped with the different corrosion behaviors and stresses that the real vessel experiences, allowing precise calculations and analyses of loads to visualise potential future outcomes for the hulk, up to and including collapse (Russell et al., 2004:37). The calculation of this model around a shipwreck that is constantly experiencing erosion and shifting weight redistributions is obviously very complex, and requires precise data based on direct measurements and observation (Russell et al., 2004:37). Corrosion analysis research on *Arizona*, therefore, has been painstakingly and regularly collected since 1983 to accurately plot the current corrosion trends of the vessel (Wilson et al., 2007:15; Murphy, 1987:10; Russell et al., 2004:36–38).

After initial non-invasive photographic and mapping surveys were completed in the early 1980s, archaeologists and conservationists not only had an inventory of what had survived the sinking event and how it was deposited in the sediment, but also the first models and drawings of the ship that researchers and the public could use as interpretive devices (Murphy, 1987:10). Sediment and water samples from within and around the vessel were taken to understand internal and external corrosive factors, visual inspections of the hull and galvanic activity studies were completed, and non-destructive hull measurements were taken, when possible, to find concretion and hull thicknesses. Minimally invasive techniques would eventually have to be used, however, for a more complete picture of the wreck's decay processes to come together. This initially included studies of biofouling composition and thickness which were supported by the in-lab study of scraped concretion material, penetrative measurement of concretion thickness, the removal of concretions from metallic hull edges, and the installation of a series of cameras along the hull to observe biochemical processes (Murphy, 1987:11–15; Henderson, 1989:117–156).

Sample pieces of the hull found in earlier salvage efforts were tested in the late 90s to understand the hull's materials (Johnson et al., 2019:1–6). Extractive techniques became necessary and sample coupons were removed from the vessel using a hydraulic-powered saw in 2002. Initially, the coupons were taken to subtract current steel thicknesses from the thicknesses illustrated on the original ship plans to develop an initial corrosion rate (Russell et al., 2004:38, 2006:312). The study of these coupons and the concretions removed in the process, however, allowed conservation scientists to develop several innovative techniques to be used in the study of concretion accumulation and corrosion potential on other shipwrecks. These have included the Concretion Equivalent Corrosion Rate (CECR) methodology, in which concretions are analyzed through x-ray diffractometry to correlate iron content in concretions with an average corrosion rate, and the Weins Number, a predictive formula that can be calculated when temperature, oxygen concentration, and concretion thickness are known (Johnson et al., 2006:55–57; Wilson et al., 2007:15–18; Johnson et al., 2011:1–7). Concretion analysis through x-ray diffractometry also assisted researchers in identifying corrosion product species and, with the environmental scanning electron microscopy of concretions, how they interact with hull metals, ultimately determining the Secant Rate of Corrosion (Wilson et al., 2007:15; Russell et al., 2006:312–317). This method of study calculated concretion growth and loss of structural mass through material transfer away from the ship hull using a referential marker left on the vessel by explosions that occurred during the sinking event. This concretion growth, along with early-stage mass loss, provides a base of comparison between *Arizona* and a linear corrosion rate previously obtained from the metal coupon samples (Johnson et al., 2018:747–751).

Along with these innovative analytical techniques, more traditional methods of marine archaeological study have been implemented in the analysis and preservation of *Arizona*. This has included interior corrosion analysis using ROVs, environmental monitoring of corrosion factors affecting the interior and exterior of the ship, structural monitoring through visual analysis and tracking, oil analysis to calculate type, breakdown, and amount of oil in the ship, study of microbiology populating the ship's hull, geological analyses of the surrounding sediments and their effects on the vessel, and the development of Geographic Information System (GIS) maps of the site, to be used for both study and public engagement (Russell et al., 2004:39–44). The results of this project are twofold: first, the FEA model, which so far has been quite accurate to the real ship in its predictions of changing stresses and visible deterioration, has provided site managers with a rough idea of the gradual rate of decay, weakest points in the hull, and possible pattern of collapse of *Arizona* moving into the near future (Foecke et al., 2010:1096–1099). Second, the project has become a model for the management and preservation of other leaking iron or steel-hulled shipwrecks, with the research potential of this effort displayed in the new techniques and prediction methods that have been developed and can be used on future shipwreck studies (Russell et al., 2004:35; Wilson et al., 2007:15–17; Johnson et al., 2011:3–6).

4.11   Conclusion

Potentially polluting wrecks link advances in our understanding of corrosion processes to events of historic and modern importance. As a result, a collateral value of shipwreck corrosion study is that it generates awareness of this most often unseen, though omnipresent, natural process and the consequences that it can have in the current environment. As site-specific biochemical and physical factors are given more time to enact their influences on shipwreck hulls, pollution events from PPWs will become a more common occurrence. Communicating this problem with decision makers, more of the scientific community, and the public will allow us to attract more concerned parties, engage it with a more thorough understanding of the processes at play and, hopefully, develop a plan to deal with the results. The longer this engagement takes, however, the more likely that catastrophic collapses will occur, transforming this generally manageable problem into an expensive and potentially deadly environmental disaster.

References

Arnold, J. B., III, Fleshman, G. M., Peterson, C. E., Stewart, W. K., Watts, G. P., & Weldon, C. P. (1992). USS monitor: Results from the 1987 season. *Historical Archaeology, 26*(4), 47–57.

Barrett, M. J. (2011). Potentially polluting shipwrecks: Spatial tools and analysis of WWII shipwrecks. Master's project. Duke University.

Beldowski, J. (2018). Dumped chemical weapons. In M. Salomon & T. Markus. (Eds,), *Handbook on marine environment protection: science, impacts and sustainable management* (Vol. 1 & 2, pp. 247–261). Springer.

Brennan, M. L., Delgado, J. P., Jozsef, A., Marx, D. E., & Bierwagen, M. (2023). Site formation processes and pollution risk mitigation of World War II oil tanker shipwrecks: Coimbra and Munger T. Ball. *Journal of Maritime Archaeology, 18*, 321–335.

Brennan, M. L., Thiemann, G., & Jeffery, W. (this volume). Satellite detection and the discovery of Bloody Marsh. In M. L. Brennan (Ed.), *Threats to our ocean heritage: Potentially polluting wrecks*. Springer.

Browne, K. (2019). "Ghost battleships" of the Pacific: Metal pirates, WWII heritage, and environmental protection. *Journal of Maritime Archaeology, 14*, 1–28.

Chan, L. S., Ng, S. L., Davis, A. M., Yim, W. W. S., & Yeung, C. H. (2001). Magnetic properties and heavy-metal contents of contaminated seabed sediments of Penny's Bay, Hong Kong. *Marine Pollution Bulletin, 42*(7), 569–583.

Chaves, I. A., Melchers, R. E., Sterjovski, Z., & Rosen, J. (2022). Long-term marine immersion corrosion of welded ABS grade steels. *Corrosion Engineering, Science and Technology, 57*(3), 195–203.

Church, R. A., Warren, D. J., & Irion, J. B. (2009). Analysis of deepwater shipwrecks in the Gulf of Mexico: Artificial reef effect of six World War 2 shipwrecks. *Oceanography, 22*(2), 50–63.

Cullimore, D. R., Pellegrino, C., & Johnston, L. (2002). *RMS Titanic* and the emergence of new concepts on consortial nature of microbial events. *Reviews of Environmental Contamination and Toxicology, 173*, 117–141.

Davin, J. J., & Witte, J. A. (1997). Cleveco underwater oil recovery: Removing a 50-year-old threat. *Proceedings of the 1997 International Oil Spill Conference, 1*, 783–788.

Delgado, J. P., Cantelas, F., Symons, L. C., Brennan, M. L., Sanders, R., Reger, E., Bergondo, D., Johnson, D. L., Marc, J., Schwemmer, R. V., Edgar, L., & MacLeod, D. (2018). Telepresence-enabled archaeological survey and identification of SS Coast Trader, Straits of Juan de Fuca, British Colombia, Canada. *Deep-Sea Research Part II, 150*, 22–29.

Duerr, R. S., Ziccardi, M. H., & Massey, J. G. (2016). Mortality during treatment: Factors affecting the survival of oiled, rehabilitated common murres (Uria Aalge). *Journal of Wildlife Diseases, 52*(3), 495–505.

Etkin, D. S., van Rooij, J. A. C., & McCay, D. F. (2009). Risk assessment modeling approach for the prioritization of oil removal operations from sunken wrecks. Effects of oil on wildlife conference. Tallin: 5-9/10/2009. Available at: https://events.sea-alarm.org/sites/default/files/14-1-rooij_full%20paper.pdf. Accessed 07/11/2023.

Evans, A. M., Firth, A., & Staniforth, M. (2009). Old and new threats to submerged cultural landscapes: Fishing, farming, and energy development. *Conservation and Management of Archaeological Sites, 11*(1), 44–54.

Eyres, D. J., & Bruce, G. J. (2012). *Ship construction: seventh edition*. Butterworth-Heinemann.

Firth, A. (2018). *Managing shipwrecks*. Fjordr Limited for Honor Frost Foundation.

Foecke, T., Ma, L., Russell, M. A., Conlin, D. L., & Murphy, L. E. (2010). Investigating archaeological site formation processes on the battleship USS Arizona using finite element analysis. *Journal of Archaeological Science, 37*, 1090–1101.

Galvele, J. R. (1983). Pitting corrosion. In J. C. Scully (Ed.), *Treatise on materials science and technology volume 23: Corrosion: Aqueous processes and passive films* (pp. 1–57). Academic.

Gilbert, T., Nawadra, S., Tafileichig, A., & Yinug, L. (2003). Response to an oil spill from a sunken WWII oil tanker in Yap State, Micronesia. *International Oil Spill Conference Proceedings 2003, 1*, 175–182.

Guedes Soares, C., Garbatov, Y., & Zayed, A. (2011). Effect of environmental factors on steel plate corrosion under marine immersion conditions. *Corrosion Engineering Science and Technology, 46*(4), 524–541.

Hac, B. (2018). Retrieval activities on the *Franken* shipwreck. *Bulletin of the Maritime Institute in Gdansk, 33*(1), 172–177.

Hamdan, L. J., Salerno, J. L., Reed, A., Joye, S. B., & Damour, M. (2018). The impact of the *Deepwater Horizon* blowout on historic shipwreck-associated sediment microbiomes in the northern Gulf of Mexico. *Scientific Reports, 8*, 9057.

Henderson, S. (1989). Biofouling and corrosion study. In D. J. Lenihan (Ed.), *Submerged cultural resources study: USS Arizona memorial and Pearl Harbor national historic landmark* (pp. 117–156). Southwest Cultural Resources Center.

Henderson, J. (2019). Oceans without history? Marine cultural heritage and the sustainable development agenda. *Sustainability, 11*, 1–22.

Jiminez, C., Andreou, V., Evriviadou, M., Munkes, B., Hadjioannou, L., Petrou, A., & Alhaija, R. A. (2017). Epibenthic communities associated with unintentional artificial reefs (modern shipwrecks) under contrasting regimes of nutrients in the Levantine Sea (Cyprus and Lebanon). *Public Library of Science (PLoS) ONE, 12*(8), 1–16.

Johnson, D. L., Wilson, B. M., Carr, J. D., Russell, M. A., Murphy, L. E., & Conlin, D. L. (2006). Corrosion of steel shipwrecks in the marine environment: USS *Arizona*—Part 2. *Materials Performance, 45*(11), 54–57.

Johnson, D. L., Carr, J. D., Wilson, B. M., Murphy, L. E., & Delgado, J. P. (2010). Corrosion of Civil War era *Sub Marine Explorer*—Part 1. *Materials Performance, 49*(9), 56–60.

Johnson, D. L., Medlin, D. J., Murphy, L. E., Carr, J. D., & Conlin, D. L. (2011). Corrosion rate trajectories of concreted iron and steel shipwrecks and structures in seawater—The Weins number. *Faculty Publications --- Chemistry Department, 197*, 1–10.

Johnson, D. L., Deangelis, R. J., Medlin, D. J., Johnson, J. E., Carr, J. D., & Conlin, D. L. (2018). The secant rate of corrosion: Correlating observations of the USS Arizona submerged in Pearl Harbor. *The Journal of the Minerals, Metals & Materials Society, 70*(5), 748–752.

Johnson, D. L., Weins, W. N., Makinson, J. D., & Martinez, D. A. (2019). Metallographic studies of the U.S.S. Monitor. In L. Berardinis, S. D. Henry, K. Marken, & M. Tramble (Eds.), *ASM failure analysis case histories: Offshore, shipbuilding, and marine equipment* (pp. 1–7). ASM.

Karlsdottir, S. N. (2022). Corrosion, scaling, and material selection in geothermal power production. In T. M. Letcher (Ed.), *Comprehensive renewable energy: Volume 7* (pp. 256–277). Elsevier.

Kery, S. M., & Stauffer, J. (2015). Hydrodynamics related to shipwreck taphonomy. *Proceedings of OCEANS 2015—MTS/IEEE Washington, 1*, 1–21.

Komai, K. (2003). Corrosion fatigue. In I. Milne, R. O. Ritchie, & B. Karihallo (Eds.), *Comprehensive structural integrity, Volume 4* (pp. 345–358). Elsevier.

Kruger, J., & Rhyne, K. (1982). Current understanding of pitting and crevice corrosion and its application to test methods for determining the corrosion susceptibility of nuclear waste metallic containers. *Nuclear and Chemical Waste Management, 3*, 205–227.

Kuroda, T., Takai, R., Kobayashi, Y., Tanaka, Y., & Hara, S. (2008). Corrosion rate of shipwreck structural steels under the sea. *Proceedings of OCEANS 2008 - MTS/IEEE Kobe Techno-Ocean, 1*, 1–6.

Lawrence, R. W. (2008). *An overview of North Carolina shipwrecks with an emphasis on eighteenth-century vessel losses at Beaufort Inlet* (Research report and bulletin series QAR-R-08-01). State of North Carolina Department of Cultural Resources.

Li, G., Zhang, Y., Xiao, J., Song, X., Abraham, J., Cheng, L., & Zhu, J. (2019). Examining the salinity change in the upper Pacific Ocean during the Argo period. *Climate Dynamics, 53*, 6055–6074.

Liao, F., Gao, G., Zhan, P., & Wang, Y. (2022). Seasonality and trend of the global upper-ocean vertical velocity over 1998–2017. *Progress in Oceanography, 204*, 1–19.

Liddell, A., & Skelhorn, M. (2019). Deriving archaeological information from potentially-polluting wrecks. In K. Bell (Ed.), *Bridging the gap in maritime archaeology: Working with professional and public communities* (pp. 81–90). Archaeopress Publishing Ltd.

Little, B. J., & Lee, J. S. (2009). Microbiologically influenced corrosion. In C. Ley (Ed.), *Encyclopedia of chemical technology* (pp. 1–38). Wiley.

MacLeod, I. D. (1987). Conservation of corroded iron artefacts—New methods for on-site preservation and cryogenic deconcreting. *The International Journal of Nautical Archaeology and Underwater Exploration, 16*(1), 49–56.

MacLeod, I. D. (1989). The application of corrosion science to the management of maritime archaeological sites. *Bulletin of the Australian Institute for Maritime Archaeology, 13*(2), 7–16.

MacLeod, I. D. (1995). *In situ* corrosion studies on the Duart Point wreck, 1994. *The International Journal of Nautical Archaeology, 24*(1), 53–59.

MacLeod, I. D. (2002). In situ corrosion measurements and management of shipwreck sites. In C. V. Ruppe & J. F. Barstad (Eds.), *International handbook of underwater archaeology* (pp. 697–714). Springer.

MacLeod, I. D. (2016a). Corrosion products and site formation processes. In M. E. Keith (Ed.), *Site formation processes of submerged shipwrecks* (pp. 90–113). University Press of Florida.

MacLeod, I. D. (2016b). In-situ corrosion measurements of WWII shipwrecks in Chuuk Lagoon, quantification of decay mechanisms and rates of deterioration. *Frontiers in Marine Science, 3*(38), 1–10.

MacLeod, I. D. (2018). Quantifying the effects of site conditions on the long-term corrosion of bronzes on historic sites. *Australasian Journal of Maritime Archaeology, 42*, 65–74.

MacLeod, I. D. (2019). Corrosion and conservation management of the submarine HMAS AE2 (1915) in the Sea of Marmara, Turkey. *Heritage, 2*, 868–883.

MacLeod, I. D., & Steyne, H. (2011). Managing a monitor—The case of HMVS *Cerberus* in Port Phillip Bay: Integration of corrosion measurements with site management strategies. *Conservation and Management of Archaeological Sites, 13*(4), 334–361.

MacLeod, I. D., & Viduka, A. (2011). The effects of storms and diving activities on the corrosion rate across the SS *Yongola* (1911) site in the Great Barrier Reef. *AICCM Bulletin, 32*(1), 134–143.

MacLeod, I. D., Richards, V., & Beger, M. (2011). The effects of human and biological interactions on the corrosion of WWII iron shipwrecks in Chuuk Lagoon. *Proceedings of 18th
International Corrosion Congress, 1*, 1–12.

MacLeod, I. D., Selman, A., & Selman, C. (2017). Assessing the impact of typhoons on historic iron shipwrecks in Chuuk Lagoon through changes in the corrosion microenvironment.
Conservation and Management of Archaeological Sites, 19(4), 269–287.

Makhlouf, A. S. H. (2015). Intelligent stannate-based coatings of self-healing functionality for
magnesium alloys. In A. Tiwari, J. Rawlins, & L. H. Hihara (Eds.), *Intelligent coatings for
corrosion control* (pp. 537–555). Butterworth Heinemann.

Makhlouf, A. S. H., Herrera, V., & Munoz, E. (2018). Corrosion and protection of the metallic
structures in the petroleum industry due to corrosion and the techniques for protection. In
A. S. H. Makhlouf & M. Aliofkhazraei (Eds.), *Handbook of materials failure analysis: With
case studies from the construction industries* (pp. 107–122). Butterworth-Heinemann.

Mann, H. (2012). The appearance of new bacteria (titanic bacterium) and metal corrosion. In
UNESCO scientific colloquium on factors impacting underwater cultural heritage (pp. 44–50).
Royal Library of Belgium.

Masetti, G. (2012). *A geo-database for potentially polluting marine sites and associated risk index.*
Master's Thesis. University of New Hampshire.

Masetti, G., & Calder, B. (2014). Design of a standardized geo-database for risk monitoring of
potentially polluting marine sites. *Environment Systems and Decisions, 34*, 138–149.

McCarthy, M. (2000). *Iron and steamship archaeology: Success and failure on the SS Xantho.*
Kluwer Academic/Plenum Publishers.

McKay, L. (2005). An unwanted legacy of war: Sunken warships in the Pacific, the environment,
and human threat. *Asia Pacific Law Review, 13*(2), 115–146.

Melchers, R. E. (1999). Corrosion uncertainty modelling for steel structures. *Journal of
Constructional Steel Research, 52*, 3–19.

Melchers, R. E. (2003). Probabilistic models for corrosion in structural reliability assessment—
Part 2: Models based on mechanics. *Journal of Offshore Mechanics and Arctic Engineering,
125*, 272–280.

Melchers, R. E. (2005). The effect of corrosion on the structural reliability of steel offshore structures. *Corrosion Science, 47*, 2391–2410.

Melchers, R. E. (2014). Long-term immersion corrosion of steels in seawaters with elevated nutrient concentration. *Corrosion Science, 81*, 110–116.

Mestre, M., Höfer, J., Sala, M. M., & Gasol, J. M. (2020). Seasonal variation of bacterial diversity
along the marine particulate matter continuum. *Frontiers in Microbiology, 11*, 1–14.

Mischler, S., & Munoz, A. I. (2018). Tribocorrosion. In K. Wandelt (Ed.), *Encyclopedia of inter-
facial chemistry: Surface science and electrochemistry volume 6.1: Corrosion and passivation*
(pp. 504–514). Elsevier.

Moffatt, C. (2004). Methodologies for removing heavy oil as used on the SS Jacob Luckenbach
and joint international testing programs. *Marine Technology Society Journal, 38*(3), 64–71.

Moore, C. (2021). *The practicalities of managing and mapping potentially polluting shipwrecks
in the UK: Legal, social and ethical considerations.* PhD thesis. University of Southampton.

Moore, J. D., III. (2015). Long-term corrosion processes of iron and steel shipwrecks in the marine
environment: A review of current knowledge. *Journal of Maritime Archaeology, 10*, 191–204.

Moore, S. W., Ambrose, J. D., & McClure, J. C. (2014). *Ulithi's deep reefs: Preliminary ROV
observations: A research report for the people of Ulithi Atoll.* California State University
Monterey Bay.

Morcillo, M., Espada, L., de la Fuente, D., & Chico, B. (2004). Metallic corrosion of the tanker
"Prestige" in deep seawater. *Revista de Metalurgia Madrid, 40*, 122–126.

Moreto, J. A., Rodrigues, A. C., da Silva Leite, R. R., Rossi, A., da Silva, L. A., & Alves,
V. A. (2018). Effect of temperature, electrolyte composition and immersion time on the elec-
trochemical corrosion behavior of CoCrMo implant alloy exposed to physiological serum and
Hank's solution. *Materials Research, 21*(6), 1–9.

Murphy, L. (1987). Preservation at Pearl Harbor. *APT Bulletin: The Journal of Preservation Technology, 19*(1), 10–15.

Naughton, J. (1985). Blast fishing in the Pacific. *South Pacific Commission Fisheries Newsletter, 33*, 16–20.

North, N. A. (1976). Formation of coral concretions on marine iron. *The International Journal of Nautical Archaeology and Underwater Exploration, 5*(3), 253–258.

North, N. A. (1984). The role of galvanic couples in the corrosion of shipwreck metals. *The International Journal of Nautical Archaeology and Underwater Exploration, 13*(2), 133–136.

North, N. A., & MacLeod, I. D. (1987). Corrosion of metals. In E. C. Pearson (Ed.), *Corrosion of marine archaeological objects* (pp. 68–98). Butterworths.

North, N. A., & Pearson, C. (1978). Washing methods for chloride removal from marine iron artifacts. *Studies in Conservation, 23*(4), 174–186.

North, N. A., Owens, M., & Pearson, C. (1976). Thermal stability of cast and wrought marine iron. *Studies in Conservation, 21*(4), 192–197.

Nürnberger, U., Sawade, G., & Isecke, B. (2007). Degradation of prestressed concrete. In C. L. Page & M. M. Page (Eds.), *Durability of concrete and cement composites* (pp. 187–246). Woodhead Publishing Limited.

Olson, S., Jansen, M. F., Abbot, D. S., Halevy, I., & Goldblatt, C. (2022). The effect of ocean salinity on climate and its implications for Earth's habitability. *Geophysical Research Letters, 49*, 1–9.

Quinn, R. (2006). The role of scour in shipwreck site formation processes and the preservation of wreck-associated scour signatures in the sedimentary record—Evidence from seabed and subsurface data. *Journal of Archaeological Science, 33*, 1419–1432.

Rasol, R. M., Bakar, A. A., Noor, N. M., Yahaya, N., & Ismail, M. (2015). Microbiologically induced corrosion monitoring using open-circuit potential (OCP) measurements. *Solid State Phenomena, 227*, 294–297.

Ridwan, N. N. H. (2019). Vulnerability of shipwreck sites in Indonesian waters. *Current Science, 117*(10), 1623–1628.

Ridwan, N. N. H., Husrin, S., & Kusumah, G. (2014). USAT *Liberty* shipwreck site in Tulamben, Karang Asem regency, Bali is under threats. *2014 Proceedings of the 2nd Asia-Pacific Regional Conference on Underwater Cultural Heritage, Honolulu, 1*, 1–12.

Russell, M. A., Murphy, L. E., Johnson, D. L., Foecke, T. J., Morris, P. J., & Mitchell, R. (2004). Science for stewardship: Multidisciplinary research on USS *Arizona*. *Marine Technology Society Journal, 38*(3), 35–44.

Russell, M. A., Conlin, D., Murphy, L. E., Johnson, D. L., & Wilson, B. (2006). A minimum-impact method for measuring corrosion rate of steel-hulled shipwrecks in seawater. *The International Journal of Nautical Archaeology, 35*(2), 310–318.

Salazar, M., & Little, B. (2017). Review: Rusticle formation on the *RMS Titanic* and the potential influence of oceanography. *Journal of Maritime Archaeology, 12*, 25–32.

Salerno, J. L., Little, B., Lee, J., & Hamdan, L. J. (2018). Exposure to crude oil and chjemical dispersant may impact marine microbial biofilm compositions and steel corrosion. *Frontiers in Marine Science, 5*, 1–14.

Siddaiah, A., Kasar, A., Ramachandran, R., & Menezes, P. L. (2021). Introduction to tribocorrosion. In A. Siddaiah, R. Ramachandran, & P. L. Menezes (Eds.), *Tribocorrosion: Fundamentals, methods, and materials* (pp. 1–16). Academic.

Silva-Bedoya, L. M., Watkin, E., & Machuca, L. L. (2021). Deep-sea corrosion rusticles from iron-hulled shipwrecks. *Materials and Corrosion, 72*(7), 1–37.

Steyne, H., & MacLeod, I. D. (2011). *In-situ* conservation management of historic iron shipwrecks in Port Phillip Bay: A study of *J7* (1924), HMVS *Cerberus* (1926) and the *City of Lanceston* (1865). *Bulletin of the Australasian Institute for Maritime Archaeology, 35*, 67–80.

Tait, W. S. (2012). Corrosion prevention and control of chemical processing equipment. In M. Kutz (Ed.), *Handbook of environmental degradation of materials* (2nd ed., pp. 865–886). William Andrew.

United States (U.S.) Navy. (2004). U.S. Navy salvage report USS Mississinewa oil removal operations. Washington Navy Yard: Naval sea systems command. S0300-B6-RPT-010: 0910-LP-102-8809.

Valenca, S. L., dos Santos, C. P., Valenca, G. O., & Neto, O. P. (2022). Investigation of corrosive processes in the hull of a shipwreck in 1905. *Research, Society, and Development, 11*(12), 1–8.

Vargel, C. (2020). *Corrosion of aluminum*. Elsevier.

Venugopal, C. (1994). Corrosion aspects of shipwreck metals of Lakshadweep waters. In S. R. Rao (Ed.), *An integrated approach to marine archaeology: Proceedings of the fourth Indian conference on marine archaeology of Indian Ocean countries, Vishakhapatnam* (pp. 33–35). Dona Paula.

Viduka, A. (2011). Managing underwater cultural heritage: A case study of SS *Yongala*. *Historic Environment, 23*(2), 12–18.

Wheeler, A. J. (2002). Environmental controls on shipwreck preservation: The Irish context. *Journal of Archaeological Science, 29*, 1149–1159.

Wilson, B. M., Johnson, D. L., Van Tilburg, H., Russell, M. A., Murphy, L. E., Carr, J. D., De Angelis, R. J., & Conlin, D. L. (2007). Corrosion studies on the USS Arizona with application to a Japanese midget submarine. *JOM: Journal of the Minerals, Metals, and Materials Society, 59*(10), 14–18.

Woloszyk, K., & Garbatov, Y. (2022). Advances in modelling and analysis of strength of corroded ship structures. *Journal of Marine Science and Engineering, 10*(6), 1–17.

Wright, J. (2016). Maritime archaeology and climate change: An invitation. *Journal of Maritime Archaeology, 11*(3), 255–270.

Xia, J., Li, Z., Jiang, J., Wang, X., & Zhang, X. (2021). Effect of flow rates on erosion corrosion behavior of hull steel in real seawater. *International Journal of Electrochemical Science, 16*, 1–13.

Yongjun Tan, M. (2023). *Localized corrosion in complex environments*. Wiley.

Zayed, A., Garbatov, Y., & Guedes Soares, C. (2018). Corrosion degradation of ship hull steel plates accounting for local environmental conditions. *Ocean Engineering, 163*, 299–306.

Zhang, Y., Zhai, X., Guan, F., Dong, X., Sun, J., Zhang, R., Duan, J., Zhang, B., & Hou, B. (2022). Microbiologically influenced corrosion of steel in coastal surface seawater contaminated by crude oil. *NPJ Materials Degradation, 6*(35), 1–12.

Zintzen, V., Norro, A., Massin, C., & Mallefet, J. (2008). Spatial variability of epifaunal communities from artificial habitat shipwrecks in the Southern Bight of the North Sea. *Estuarine, Coastal, and Shelf Science, 76*(2), 327–344.

Open Access This chapter is licensed under the terms of the Creative Commons Attribution 4.0 International License (http://creativecommons.org/licenses/by/4.0/), which permits use, sharing, adaptation, distribution and reproduction in any medium or format, as long as you give appropriate credit to the original author(s) and the source, provide a link to the Creative Commons license and indicate if changes were made.

The images or other third party material in this chapter are included in the chapter's Creative Commons license, unless indicated otherwise in a credit line to the material. If material is not included in the chapter's Creative Commons license and your intended use is not permitted by statutory regulation or exceeds the permitted use, you will need to obtain permission directly from the copyright holder.

Chapter 5
From Desktop to Dive: Assessing the Pollution Potential of SS *Fernstream*, USNS *Mission San Miguel* and SS *Coast Trader*

James P. Delgado

5.1 Introduction

In 2010, the United States Congress appropriated one million dollars to identify the most ecologically and economically significant potentially polluting wrecks in US waters, a joint initiative between the U.S. Coast Guard and its Regional Response Teams, along with NOAA, through its Office of National Marine Sanctuaries, Office of Response and Restoration, and Maritime Heritage Program. Of some 20,000 US wrecks that postdated 1891, when most steam and motor vessels converted from coal to oil, the study focused on: vessels built of iron or steel (wooden wrecks would have deteriorated and already released their oil), cargo vessels over 1000 tons (smaller vessels would have small amounts of fuel), and any tank vessel. This resulted in a group of 600–1000 vessels.

Detailed research followed, working from available wreck reports, archival documents, and resulted in a focused desktop study of 87 vessels that were thought to pose a potential pollution threat due to the nature in which they sank, the amount of fuel or cargo likely still inside the wrecks, and the structural reduction and demolition of those that were navigational hazards. To further screen and prioritise these vessels, risk factors and scores were applied to elements such as the amount of oil that could be on board and the potential ecological or environmental impact. The contractors did the modeling forecasts, as well as ecological and environmental 'resources at risk' assessments. In nearly every evaluation of the 87 vessels, little to no details of the actual wreck as a physical entity was available that offered a forensic sense of the wreck, and few of the wrecks had even yet been located.

The NOAA study had a budget of one million (USD) set by Congress; equally divided, that means that each of the 87 studies would have cost $11,494.25. Actual

J. P. Delgado (✉)
SEARCH Inc., Washington, DC, USA
e-mail: james.delgado@searchinc.com

© The Author(s) 2024

M. L. Brennan (ed.), *Threats to Our Ocean Heritage: Potentially Polluting Wrecks*, SpringerBriefs in Underwater Archaeology,
https://doi.org/10.1007/978-3-031-57960-8_5

costs varied, of course, and additional resources were brought to bear, such as NOAA and US Coast Guard salaries, but in the larger scale of government spending, the Congressionally set budget was minimal. Within that context, a national effort to locate and conduct a physical assessment of each of the 87 wrecks, an approach which had not been funded, was beyond practical reach. While some wrecks were in shallow water, and more readily accessible, others lay deep, where the technological means to reach them for assessment was cumulatively cost-prohibitive.

The obvious solution for a 'next step' that took the project from desktop to dive would come with oceanographic cruises, surveys and remotely operated vehicle dives of opportunity. Three separate opportunities in 2013, 2015 and 2016 resulted in physical assessments of the wrecks of SS *Fernstream* (1952), SS *Coast Trader* (1942) and USNS *Mission San Miguel* (1957) and subsequent revisions to the initial evaluations of their potential to pollute. The author was one of the NOAA managers involved in the NOAA PPW study and was principal archaeological investigator for two of the reassessments as Director of NOAA's Maritime Heritage Program in the Office of National Marine Sanctuaries.

5.2   SS *Fernstream*

The Norwegian steel-hulled motor vessel *Fernstream* was built at Eriksbergs Mekaniska Verkstads Aktiebolag in Gothenburg, Sweden as one of six 8400-ton freighters built to the same plan in 1948–1949. Launched in July 1949, the 416 ft (126 m) long freighter entered service in October as a general-purpose cargo ship with refrigerated holds and accommodations for passengers. The Norwegian shipping firm A/S Glittre owned *Fernstream*, but it was managed for them by the Oslo-based firm of Fearnley & Eger under the Norwegian flag. Unlike earlier steamships, *Fernstream* and its sister ships were diesel-powered. Under Fernley & Eger's flag, *Fernstream* and its sisters shipped cargo from the United States to Asia, and in some cases, immigrants to Asia Pacific countries.

On its final voyage, *Fernstream* departed San Francisco for Manila with 11 passengers, 3000 tons (2721 mt) of soybeans in bulk, and general cargo that included machinery for a hydro-electric power plant in the Philippines on December 11, 1952. It was foggy, and as *Fernstream* was in position to motor out of the Golden Gate, the inbound cargo freighter *Hawaiian Rancher*, an 8353-ton vessel, struck *Fernstream* on the port side near the bridge, penetrating the hull and flooding the after part of the engine room. The collision damaged the watertight bulkhead connecting the engine room to the #4 hold, and the crew was unable to close the watertight doors in the shaft alley. Flooding rapidly, and with a total loss of power, *Fernstream* sank quickly as the damaged but still afloat *Hawaiian Rancher* lowered lifeboats and rescued *Fernstream*'s passengers and crew.

Fernstream's wreck was not salvaged nor was there any historical record of efforts to clear or reduce the profile of the wreck. When evaluated during the NOAA

PPW study, *Fernstream* was classified as a moderate/high risk of polluting San Francisco Bay and the surrounding coastline, depending on tides and the transfer of water in and out of the Golden Gate, because there were no accounts of clearance, or of oil release at the time of the sinking. However, the study noted that there were no detailed reports of the sinking, other than local news headlines, and those focused on the human drama of the rescue of the passengers and crew. It was also assumed that the wreck was likely not broken up and was in one contiguous piece:

> The *Fernstream* is classified as High Risk for oiling probability for shoreline ecological resources for the WCD because 100% of the model runs resulted in shorelines affected above the threshold of 100 g/m². It is classified as Medium Risk for degree of oiling because the mean weighted length of shoreline contaminated was 52 miles. The *Fernstream* is classified as High Risk for oiling probability to shoreline ecological resources for the Most Probable Discharge because 100% of the model runs resulted in shorelines affected above the threshold of 100 g/m². It is classified as Medium Risk for degree of oiling because the mean weighted length of shoreline contaminated was 18 miles (NOAA, 2013b: 27).

The prominent location of the wreck, just off Alcatraz in one of the most scenic harbors of the United States, while not a factor in the assessment of potential risk, did result in NOAA recommending further assessment of the wreck, using surveys of opportunity to locate and assess it.

The opportunity to do so came just as the PPW study was released by NOAA. A multi-year study of maritime heritage resources in Greater Farallones National Marine Sanctuary, headquartered in San Francisco, involved multiple partners (Delgado et al., 2020). In May 2013, NOAA's Office of Coast Survey's locally based Navigation Response Team (NRT 6) conducted two side-scan sonar surveys that relocated and mapped the wreck. The demonstration of a CODA Octopus Echo-Scope sonar system for the dive team of the San Francisco Fire Department's Marine Unit in November 2013 also assessed *Fernstream* on NOAA's recommendation. The sonar data from the three surveys was highly illustrative. *Fernstream* sits upright with the forecastle, partial remains of the stern-house and bridge-house, as well as the forward and after mast house structures visible. The hull has suffered from catastrophic collapse of the bridge house structure. The hull is breached on the starboard side forward of the bridge-house, the location of the collision. Masts, booms, and king posts have collapsed onto the deck; and the port side of the wreck is buried deep in sediment.

The stern is the highest remaining portion of the wreck above the seafloor but was seen to be partially collapsed. The bow has twisted to port, and the bow is no longer in longitudinal alignment with the after part of the hull forward of the bridge house. Outside the wreck, the starboard side of the hull shows evidence of sediment scouring, more prominently in the bow, a typical occurrence with shipwrecks due to their position on the seafloor in prevailing currents. Rather than a contiguous wreck, *Fernstream* is broken, and heavily shrouded in the thick muds of the bay. The deep tanks, where fuel oil was bunkered, are at the lowest levels of the hull, all of which is now buried deeply in bay mud.

Based on the sonar imagery, NOAA revised the PPW assessment for *Fernstream*, noting that the sonar indicated that *Hawaiian Rancher* appears to have impacted more than one deep tank containing fuel oil, and either from settling on the seabed or after decades of corrosion in the active currents of San Francisco Bay near the harbor entrance, *Fernstream*'s starboard hull has suffered a catastrophic collapse that likely also collapsed the #3 deep tank. The hull collapsed and is open to the sea where exposed, and likely no longer contains any significant amounts of oil, but the possibility of oil in sediment filled areas of the wreck remains a probability. The risk of the wreck polluting was modified from high to medium, with a recommendation for active monitoring if sediment movement released oil at a future date.

5.3   USNS *Mission San Miguel*

The USNS *Mission San Miguel* was a 524 ft (160 m) long T2-SE-A2 tanker, with a welded steel hull and turbo-electric propulsion. It was one of more than 500 mass-produced T2 tankers built to the design specifications of the U.S. Maritime Commission during the Second World War (see Brennan et al., Chap. 9, this volume). Built in 1943–1944 at the Marinship Yard in Sausalito, California for the War Shipping Administration, it was chartered to private owners until 1946, when it was laid up at the National Defense Reserve Fleet moorage near Mobile, Alabama. Acquired by the U.S. Navy in 1947 and commissioned as USS *Mission San Miguel* (AO-129) it was chartered again to private owners, before being transferred to the Military Sea Transportation Service as USNS *Mission San Miguel*. The tanker passed in and out of service with extended layups in reserve fleets through 1957, when it wrecked on October 8, 1957, after leaving Guam and heading across the Pacific. It ran aground during rain squalls on Maro Reef in the Northwestern Hawaiian Islands. Fully ballasted with sea water, it remained intact, and the crew was rescued (NOAA, 2013c).

The historical record, which includes US Navy salvage reports and damage assessment from the unsuccessful attempt to recover the vessel at the time of its wrecking on Maro Reef, along with a detailed assessment of the record by archaeologist Hans Van Tilburg, makes it clear that the tanker was in ballast at the time of loss, with no cargo (fuel oil and gasoline) on board (Van Tilburg, 2003). USNS *Mission San Miguel* was fueled and after running for 5 days had consumed an estimated 1700 barrels of Navy Special Fuel (No. 5) out of a 14,700-barrel capacity. The fuel oil was carried in two tanks, which were located aft of the engineering compartment bulkhead.

The tanker could not be freed and was abandoned to break up in that remote archipelago of atolls and reefs. The area of the wreck is now within Papahānaumokuākea Marine National Monument, primarily administered by NOAA in partnership with other agencies. The presence of the wreck in the marine protected area led to an assessment of potential risk in the 2013 study even though the wreck had yet to be located (Van Tilburg, 2002, 2003). Given the uncertainty of how much of the tanker's fuel remained inside the wreck, it was rated a High/Medium, which at the time was noted as a conservative assessment reflecting the lack of information on both exact location and structural integrity. As annual Rapid Ecological Assessment and Monitoring (RAMP) cruises included maritime heritage components, the hope was that one of these cruises would locate and document *Mission San Miguel*.

That opportunity came with a RAMP cruise in August 2015. The wreck was located on August 3, 2015, and a preliminary archaeological assessment was undertaken by the team of archaeologists and scientists aboard the NOAA vessel *Hi'ialakai*. The vessel lies close to the reported area of its loss, in 80 ft of water off Maro Reef. An archaeological team documented the site and prepared a site drawing and a site report (Raupp, 2015) The 2015 assessment focused on 2 days of dives to document the characteristics of the site and compare those data with plans of USNS *Mission San Miguel* to identify the wreck and to assess its archaeological integrity as a site potentially eligible for listing in the National Register of Historic Places.

The wreck lies at a bearing of 080° and consists of three main sections. These include the largely intact stern section of the ship, which rests on its port side and rises from approximately 80 ft (24 m) at the seabed to a depth of 29 ft (8.8 m) below the water surface; the aft cargo tank section, which is located directly forward of the stern and is mostly collapsed; and an area thought to represent the tanker's bow located approximately 540 ft (165 m) east and northeast of the stern section that includes two anchors and chain, numerous components of deck machinery, and scattered sections of hull plate and piping in depths ranging from 5 to 35 ft (1.5–10.5 m). Each of these aspects of the vessel's deposition closely matches historical data pertaining to *Mission San Miguel*. According to US Navy reports, when salvage operations were terminated the tanker was aground with a bearing of 075° and drew 4 ft in the forward part and 83 ft in the after portion (Raupp, 2015).

The 2015 survey documented the disintegration of the majority of the hull of the vessel and that this had extended into the engineering spaces in the stern. As the vessel sank and moved deeper, it is likely that the tanker twisted and broke apart. The bow, being the shallowest portion of the wreck, has collapsed and broken apart. Only the stern is intact. Structural collapse of all but approximately 100 ft (33 m) or a fifth of the hull includes the fuel oil tank on the starboard side of the stern. About 18 ft (5.5 m) of the port side and therefore the port fuel oil tank is buried in the sand bottom. The archaeological team believes that, while this tank may not be collapsed, it may have opened up to some extent due to the pressure of the wreck. They observed no oil or sheening on the site. Divers did not enter the wreck but visible overhead spaces did not contain any obvious oil (Raupp, 2015).

When assessing the archaeological report, NOAA revised the risk assessment for the wreck. The Pollution Potential score for the vessel is now LOW for worst case discharge and LOW for most probable discharge. This downgrade was based on the change in knowledge of the wreck site, particularly the lack of remaining structural integrity in the areas of the wreck that would have held fuel oil. No further work was recommended as the only other archaeological work that could be undertaken would be excavation of the buried port tank area but that could have released trapped oil.

The existence of oil residue or trapped oil beneath portions of the intact stern is possible, but NOAA managers believe the amounts would be minimal.

5.4   SS *Coast Trader*

The freighter *Holyoke Bridge* was built during the First World War by the Submarine Boat Company at Newark, New Jersey as part of the United States' response to the war and the need for ships due to losses to U-boats. In the 2 weeks following the United States' entry into the war in April 1917, German submarines sank 122 ships as part of a calculated strategy of unrestricted warfare. The American response was to declare an emergency and create the U.S. government subsidised Emergency Fleet Corporation (EFC), which the U.S. Shipping Board established in April 1917 to meet the emergency need for more ships (Hurley, 1927). Hundreds of vessels—steel, concrete and wooden-hulled—were laid down and completed during the remaining year of the war in Europe and up into the early 1920s.

Holyoke Bridge, laid down in late 1919, was launched at the end of January 1920. It was a riveted steel-hulled steamship, 324 ft (98.8 m) long, with open cargo holds for bulk commodities or crated goods, and its fuel capacity was 8088 barrels of Bunker C heavy fuel oil. Owned by the U.S. Shipping Board, it was chartered to private companies, all part of an expanded postwar U.S. merchant fleet that worked coastal routes as well as more extended voyages to Central and South America, ports in the Pacific, and transatlantic destinations. During the 1920s and 1930s, renamed SS *Point Reyes*, and then *Coast Trader*, the steamer worked out of the Gulf of Mexico and along the U.S. west coast until the outbreak of the Second World War. Chartered to the U.S. Government's War Shipping Administration, *Coast Trader* was lost to Japanese submarine attack on June 7, 1942, after it departed Port Angeles, Washington with a cargo of newsprint, bound for San Francisco.

When off the coast of Vancouver Island and past the Straits of Juan de Fuca, *Coast Trader* was hit with a single torpedo near the stern by the Imperial Japanese Navy submarine I-26; the 56 members of the crew were able to escape the sinking without fatalities as *Coast Trader* sank by the stern, disappearing 40 min after the explosion. The position of the loss was generally known but not charted, and at the time of the PPW study, *Coast Trader* was, while never precisely located as a shipwreck, thought to be on the maritime border of the U.S. and Canada and off the coast of Washington's Olympic Peninsula and British Columbia's Vancouver Island. While it might have still held a full capacity fuel load of 8088 barrels (339,696 gallons/1,286,889 liters) it was rated a MEDIUM risk:

> For the Worst Case Discharge, *Coast Trader* scores Low with 11 points; for the Most Probable Discharge (10% of the Worse Case volume), *Coast Trader* also scores Low with 10 points…survivor reports of the sinking make it sound like substantial amounts of oil was lost when the vessel sank, it is not possible to determine with any degree of accuracy what the current condition of the wreck is and how likely the vessel is to contain oil since the shipwreck has never been discovered (NOAA, 2013a:5).

The study also noted that 'based on the large degree of inaccuracy in the reported sinking location and the depths of water the ship was lost in, it is unlikely that the shipwreck will be intentionally located' (NOAA, 2013a:5, 34).

At that time, what was not known was that a Canadian Hydrographic Service multibeam sonar survey on October 15, 2010, imaged a shipwreck off the entrance to the Strait of Juan de Fuca, which would prove to be *Coast Trader*. Jacques Marc of the Underwater Archaeological Society of British Columbia, a regional nonprofit avocational archaeology group that works in close coordination with the Province of British Columbia's Cultural Branch, assessed the sonar data and concluded the wreck was likely *Coast Trader*. While this was not known to NOAA at the time of the 2013 report, sharing of data with the Canadian Coast Guard and UASBC led to a decision to deploy an ROV to assess it on a NOAA-funded expedition to the region by the Ocean Exploration Trust vessel E/V *Nautilus* that would work both in U.S. and Canadian waters in June–July 2016.

E/V *Nautilus* made a single extended ROV dive on the sonar target on June 2, 2016. A team from NOAA, OET, and U.S. Coast Guard, the Underwater Archaeological Society of British Columbia and the Vancouver Maritime Museum joined via remote telepresence from the University of Rhode Island-based Inner Space Center (Delgado et al., 2018). The wreck lay in 541 ft (165 m) off Victoria, British Columbia. The multi-hour dive verified that the wreck was *Coast Trader*, which at that time had not been visually inspected and remained officially unidentified.

The wreck was found on its keel, upright on the seabed, oriented on a 346-degree heading. The torpedo detonation tore off the stern and essentially destroyed the freighter from just aft of its midships deckhouse. The bow was intact, and this high degree of structural preservation enabled the team to quickly match features on the wreck to the plans and images of *Coast Trader*. The forensic inspection of the wreck found that the reported area of the torpedo hit was slightly off; instead of hitting at the area of Hold #4, it hit Hold #3, blowing off the stern, which lay attached by torn steel plating to the intact forward half of the wreck. If fully fueled, *Coast Trader* lost half of its presumed oil at the time of the torpedo attack. The remaining likely remains inside the well-preserved, lightly corroded forward half.

While the wreck's lighter steel superstructure and decks are corroded and have failed in places, the hull appears intact and without visible corrosion. The key question is whether *Coast Trader* will soon collapse and catastrophically release whatever oil remains as opposed to a slower release of oil through vents and piping over time. Based on the visual inspection, the 2016 team did not believe this was either imminent or possible for a number of years. The result of the survey was an archaeological report, but also a revision of the PPW score for *Coast Trader*. As noted at the time, the dive on *Coast Trader* added to our understanding of the event provided the means by which a more detailed assessment of the wreck as both a historic site and a potential pollution hazard could be completed. This demonstrated that while wrecks with fuel left inside are a concern, some, like *Coast Trader*, do not need expensive mitigation for the foreseeable future.

5.5   Conclusion

Desktop assessments have value, especially when a task as daunting as assessing the pollution potential for tens of thousands of shipwrecks is requested. The ultimate selection of 87 vessels as the highest-risk potentially polluting shipwrecks in U.S. waters substantially narrowed the focus, but the work was still considerable. In some cases, insufficient archival information was available, and while some shipwrecks were 'known,' insufficient detail of each wreck's position, damage, structural integrity and change over time was lacking. As noted, detailed assessment would only come with cruises or dives of opportunity. The cruises and investigations 'of opportunity' model proved effective, both in assessing risk and also the site characteristics of historic shipwrecks. At the same time, other assessments have found large amounts of oil and remediation was necessary, as in the case of SS *Coimbra* and SS *Munger T. Ball* (Brennan et al., 2023), and is most likely with SS *Bloody Marsh* (Brennan et al., Chap. 9, this volume).

In the three cases presented here, valuable information was obtained. *Coast Trader* was found to not be in U.S. waters, but in Canada's. This did not change the assessment in regard to areas of impact, which had primarily been scientifically calculated to be in Canada, but the determination that at least half of the potentially spillable fuel had been discharged in 1942, and the structural integrity of the surviving forward area of the wreck did shift the assessment of risk. Similarly, the realisation that the sinkings and post-loss changes to the wrecks of both *Fernstream* and *Mission San Miguel* had also released oil, likely in small amounts over a protracted period, which lessened the risk of a modern major event. The obvious conclusion is that opportunities to assess potentially polluting, historic shipwrecks should be a priority in planning for research cruises. This aspect is now being applied in the United States by NOAA's Office of Ocean Exploration as a result of the assessments of *Fernstream*, *Mission San Miguel* and *Coast Trader*.

References

Brennan, M. L., Delgado, J. P., Jozsef, A., Marx, D. E., & Bierwagen, M. (2023). Site formation processes and pollution risk mitigation of World War II oil tanker shipwrecks: *Coimbra* and *Munger T. Ball*. *Journal of Maritime Archaeology, 18*, 321–335.

Delgado, J. P., Cantelas, F., Symons, L. C., Brennan, M. L., Sanders, R., Reger, E., Bergondo, D., Johnson, D. L., Marc, J., Schwemmer, R. V., Edgar, L., & MacLeod, D. (2018). Telepresence-enabled archaeological survey and identification of SS Coast Trader, Straits of Juan de Fuca, British Colombia, Canada. *Deep-Sea Research Part II, 150*, 22–29.

Delgado, J. P., Schwemmer, R. V., & Brennan, M. L. (2020). Shipwrecks and the maritime cultural landscape of the Gulf of the Farallones. *Journal of Maritime Archaeology, 15*, 131–163.

Hurley, E. N. (1927). *The bridge to France*. The J.B. Lippincott Company.

NOAA. (2013a). *Screening level risk assessment package*. Coast Trader.

NOAA. (2013b). *Screening level risk assessment package*. Fernstream.

NOAA. (2013c). *Screening level risk assessment package*. Mission San Miguel.

Raupp, J. (2015). Discovery of USNS Mission San Miguel (T-AO-129), ex-Mission San Miguel. Report submitted to NOAA's Office of National Marine Sanctuaries. Manuscript on file, NOAA headquarters, Silver Spring, MD.

Van Tilburg, H. K. (2002). Maritime cultural resources survey: Northwestern Hawaiian Islands NOWRAMP 2002. Report submitted to National Oceanic and Atmospheric Administration National Ocean Service NWHICRER.

Van Tilburg, H. K. (2003). *US Navy shipwrecks in Hawaiian waters: An Inventory of submerged naval properties*. Naval Historical Center.

Open Access This chapter is licensed under the terms of the Creative Commons Attribution 4.0 International License (http://creativecommons.org/licenses/by/4.0/), which permits use, sharing, adaptation, distribution and reproduction in any medium or format, as long as you give appropriate credit to the original author(s) and the source, provide a link to the Creative Commons license and indicate if changes were made.

The images or other third party material in this chapter are included in the chapter's Creative Commons license, unless indicated otherwise in a credit line to the material. If material is not included in the chapter's Creative Commons license and your intended use is not permitted by statutory regulation or exceeds the permitted use, you will need to obtain permission directly from the copyright holder.

Chapter 6
Managing Potentially Polluting Wrecks in the United Kingdom

Polly Georgiana Hill, Matthew Skelhorn, and Freya Goodsir

6.1 The Inception of Wreck Management in the UK

The battleship HMS *Royal Oak* was at anchor in Scapa Flow in Orkney when, in the early hours of 14th October 1939, the German submarine U-47 entered the harbour and fired a salvo of torpedoes at *Royal Oak's* port side. U-47 then repositioned itself at close range and fired a further three torpedoes at the starboard side of the ship. The weather had been fair, so all the ship's hatches were open and consequently it took on water very quickly and sank in just 13 min with the loss of 833 of its crew.

HMS *Royal Oak* lies partially inverted on its starboard side in approximately 30 m of water, with the shallowest part only 3 or 4 m below the sea surface. The ship was fully loaded with 3500 m³ of diesel and heavy fuel oil when it was attacked. An estimated 485–735 tonnes of oil were released during the sinking, and more was washed out during the subsequent war years. This leakage stopped by about 1945, but a gradual corrosion of rivets and seams eventually led to a slow but increasing leak of fuel from the wreck from about 1960. By the mid-1990s, disquiet about the increasing leakage, together with growing concern about the stability of the inverted wreck and the potential for a significant sudden release of oil, led to pressure for something to be done to protect the Scapa Flow environment.

In February 1995, the Secretary of State for Defence accepted that the UK Ministry of Defence had a moral responsibility to intervene as the owners of the wreck. In October 1996 a patch was placed over the area of worst leakage as a short-term measure while longer-term options were explored. In January 1999 an oil-collection canopy was fitted over the worst leak, but it was ripped off in storms a

___

P. G. Hill (✉) · M. Skelhorn
Salvage and Marine Operations, Ministry of Defence, Bristol, UK
e-mail: polly.hill@ipieca.org; Mathew.Skelhorn219@mod.gov.uk

F. Goodsir
Centre for Environment, Fisheries and Aquaculture Science, Lowestoft, UK

© The Author(s) 2024
M. L. Brennan (ed.), *Threats to Our Ocean Heritage: Potentially Polluting Wrecks*, SpringerBriefs in Underwater Archaeology,
https://doi.org/10.1007/978-3-031-57960-8_6

71

couple of weeks later, taking the patch with it. At this stage it was accepted that the best option was to remove the remaining fuel on board.

The Ministry of Defence turned to its Salvage and Marine Operations team. In 2001, a team of Salvage and Marine Operations engineering divers deployed on a pilot operation to prove the practicality of hot-tapping the tanks of HMS *Royal Oak*. Hot tapping involves attaching a valve to the tank boundary and hollow-cutting a hole concentric with the valve, making sure that the boundary is sealed throughout, so that oil isn't released. A hose can then be attached to the valve and the tank contents pumped out into another vessel. The wreck was boomed to collect any escaped oil. The pilot was successful, and a full hot-tapping operation was planned to remove the remaining oil.

The oil pollution from HMS *Royal Oak* forced the Ministry of Defence to face the environmental legacy of its modern naval warfare and the Wreck Management Programme was born in 2008 to proactively manage the environmental and safety risks associated with its remaining wreck inventory.

6.2   Current Wreck Management in the UK

UK-owned Potentially Polluting Wrecks (PPWs) are now actively managed between two different government departments, and a few other government departments also have an interest. Ships that were in the service of the Royal Navy or the Royal Fleet Auxiliary at the time of sinking are the responsibility of the Ministry of Defence. This includes ships that were taken up from industry for the war effort in World War I (WWI) and World War II (WWII), including tankers that were transporting fuel oil under the flag of the Royal Fleet Auxiliary. Ministry of Defence wrecks are managed by two teams, one within Navy Command who manage heritage issues for all Ministry of Defence wrecks, and one within Salvage and Marine Operations who manage the safety and environmental risks associated with wrecks through the Wreck Management Programme. The Ministry of Defence is responsible for an estimated 5600 PPW globally.

The Department for Transport has policy responsibility for approximately 5300 mainly merchant shipwrecks, which includes some PPWs. These came into government ownership largely through the War Risk Insurance scheme which provided payments to their owners for any loss due to enemy action during WWI and WWII. Depending on the nature of the insurance policy, the government then took ownership of the hull and/or cargo of the resulting wreck. There are, however, some wrecks where more than one government department may have an interest. For example, a preliminary assessment suggests that the Department for Transport owns the hull of the wrecked SS *Derbent,* which sank holding a cargo of Ministry of Defence-owned oil.

These UK government teams have a global remit, excluding wrecks in US waters. This is due to a 'knock for knock' agreement between the UK and the US which states that each government will waive all claims arising from or connected with negligent navigation in respect to any cargo (Provision of Mutual Aid Concerning Certain Problems of Marine Transportation and Litigation Agreement, 1942). This has enabled the UK and US governments to take responsibility for historic wreck risk management in their respective waters, which has been of mutual benefit. For example, problematic British wrecks such as the tanker SS *Coimbra* that sank off Long Island in 1942 leaked oil for decades until a project coordinated by the US Coast Guard hot tapped the wreck and removed an estimated 500,000 gallons of oil (Brennan et al., 2023). Conversely, the UK government has managed the US-owned SS *Richard Montgomery* located in the Thames Estuary.

The Ministry of Defence Wreck Management Programme aims to identify wrecks that still contain significant quantities of oil so that this can be removed in a controlled operation, as was done with HMS *Royal Oak*. Although some wrecks contain other hazardous materials, and all contain at least some munitions, oil has hitherto been the priority due to the potential volumes involved. The process begins with the prioritisation of the Ministry of Defence's entire post-1870 (when ships began the transition to fuel oil) shipwreck inventory based on criteria such as ship size and proximity to shore (Liddell & Skelhorn, 2018). This prioritisation, using easily accessible information, identified several hundred high priority wrecks including tankers and large warships lost with significant quantities of oil. Following prioritisation, several wrecks have been investigated further to identify and mitigate the environmental risks they pose.

The Department for Transport is building upon the work of the Ministry of Defence Wreck Management Programme and, following a desk-based review, continues to assess its wrecks in line with the standardised approach to the environmental risk assessment of potentially polluting wrecks (Goodsir et al., 2019). It has also made its wreck data available to develop innovative methods for identifying historic pollution, such as the Department for Environment, Food and Rural Affairs (Defra) Earth Observation Centre of Excellence project which aims to develop a tool for mapping historic oil spills using satellite data.

6.3 Distribution of British PPW

The distribution of Ministry of Defence wrecks mirrors the UK's involvement in WWI and WWII. The majority are concentrated in northern European waters with significant numbers in the Mediterranean and, to a lesser extent, the Far East and Baltic Sea. Patterns are discernible within these regional groupings. For example, the convoy routes across the Atlantic Ocean and round to the northern Russian ports can be picked out from the trail of wrecks left behind, while other concentrations denote campaigns or individual battles. A significant grouping of wrecks marks the UK's involvement in the Gallipoli campaign, for example, while the ebb and flow of the Battle of Jutland can be charted from the remains of the vessels sunk during that engagement.

The distribution of WWI UK merchant vessel wrecks reflects the enormous efforts of the Mercantile Marine to supply the country and its armies during the conflict. While this effort is mirrored in their geographical spread, the distribution also reveals significant temporal variation reflecting the way the war was conducted. For example, prior to the general introduction of the convoy system in 1917, the merchant fleet was subject to huge losses as individual ships were picked off by marauding U-boats. The loss rate fell significantly after the introduction of convoys.

While many of the wrecks remain to be discovered, the spread of WWII UK Merchant Navy losses again mirrors the strategic and tactical decisions that framed the conflict. The convoy system was initiated from the outset and the efforts of the axis powers to counter it can be discerned from the groups of merchant losses marking successful U-boat wolfpack, warship and aircraft attacks on individual convoys. Consequently, the day-to-day interplay between the axis and allied powers has implications at the present. The individual running convoy battles that frequently developed are marked by temporally, and sometimes geographically, confined groups of wrecks that may pose environmental and safety risks at the present day.

Although UK wrecks are spread around the world, they have not received equal treatment with regards the management of environmental and safety concerns. Most work has been concentrated within UK territorial waters. Further afield efforts have been patchier. In part this reflects the sheer volume of work close to home coupled with the costs and difficulties of operating more remotely. However, it also reflects the fact that an appreciation of the potential risks posed by legacy wrecks, and associated demand for action, varies between countries. For example, Norway has pursued an extremely proactive policy of oil removal from foreign legacy wrecks within its waters (Bergstrøm, 2014). Conversely, relatively little work on PPWs has taken place in the Mediterranean Sea despite the large number of losses that occurred there in both World Wars. However, this situation is likely to change with the first signs of initiatives to better understand the risks posed by Mediterranean wrecks.

6.4   Wreck Survey Techniques

Archival research is a useful starting point to determine what fuel and cargo a wreck might have sunk with, how much damage was caused at the time of sinking, and where the wreck now lies (Liddell & Skelhorn, 2018), but to minimise the assumptions in a risk assessment, the wrecks need to be surveyed (Hill et al., 2022). The same techniques are generally used, but not all wrecks are the same so the survey approach may vary from wreck to wreck. The chosen survey method(s) depend on how much (or little) is already known about the wreck and therefore the level of concern, the wreck location and depth, and the type of ship, its size and construction.

Survey techniques typically used to assess the structural integrity of wrecks and their tanks include comparison of ship plans with multibeam echosounder (MBES) and live feed video surveys using Remotely Operated Vehicles (ROVs) and towed side-scan sonar surveys. Where wrecks are surveyed over multiple years it is possible to follow the collapse (or illegal salvage) of the wreck over time. The deterioration of wrecks can also be detected by satellite imagery of oil spills. Advances in technology mean survey techniques are likely to evolve in the next few years. An increased use of autonomous vehicles to survey wrecks will reduce the carbon footprint of the Wreck Management Programme and increase its output.

Environmental surveys have been completed for some priority wrecks to measure concentrations of oil derived contamination within sediment, biota and the water column where historical and active leaks have occurred (e.g., Hill et al., 2022). Sediment is typically collected by grab or corer, and biological samples are collected by beam trawls; the type of equipment deployed is influenced by the sediment type, hydrodynamic conditions, and other hazards in the area. Water fluorescence (indicative of hydrocarbon concentration) is measured in real time and water samples are collected and analysed to convert fluorescence data to hydrocarbon concentration.

Oil spill modelling with integrated wind and current data can predict the transport and fate of oil in the marine environment, indicating likely areas of elevated seawater and sediment contamination. Survey design is influenced by model outputs, MBES bathymetry, and other anecdotal evidence such as prevailing seabed currents. Sample points are orientated along radiating transects centred on the wreck, with the primary focus on the most probable direction of travel of released oil to increase the likelihood of detecting contamination. Control samples may be collected away from the wreck for comparison.

Historic leaks can be detected in surface or sub-surface sediment layers because oil persists in marine sediments for many years. Fine sediments such as muds and clays have a higher tendency to accumulate Polycyclic Aromatic Hydrocarbons (PAHs), particularly in non-mobile sediments with low oxygen levels, and can be targeted to assess contaminant load and the potential impact of sediment resuspension. Total Hydrocarbon Content (THC) and summed PAH (ΣPAH) are quantified using coupled gas chromatography-mass spectrometry using accredited methods (Kelly et al., 2000). Elevated concentrations can pose a risk to organisms living or feeding in the region and reduce biodiversity there. Characterisation of the physical habitat and local ecology can be used to measure ecological health (Thomas et al., 2021) and provide a baseline of sensitive receptors for comparison with subsequent monitoring.

The Wreck Management Programme compliments its own surveys with other sources of information, predominantly from academia. For example, Bangor University shared MBES and dive footage of the tanker SS *Derbent*, which suggested the wreck was intact and required further investigation (see case study below). An equally useful resource is the diving community, who are passionate about wrecks and are generally committed to diving responsibly and respectfully. Community involvement has proved an effective way of documenting wrecks (Viduka, 2020) and given appropriate training, this can be extended to the environmental and safety management of wrecks.

6.5   Case Studies

A selection of case studies is presented to demonstrate the diversity of risks and challenges that wrecks present, and the action the UK government has taken to assess and, occasionally, mitigate the risks they pose.

6.5.1   *RFA* Darkdale

The tanker RFA *Darkdale* was stationed at the South Atlantic island of St Helena acting as a refuelling tanker for Royal Navy vessels when it was torpedoed and sunk by *U-68* on 22 October 1941. The wreck of the RFA *Darkdale* lies in James Bay, approximately 600 m offshore from Jamestown, which is the capital of the island, at around 45 m depth. The wreck had sporadically leaked small quantities of oil but a storm in the spring of 2010 led to a significant release from the wreck. St Helena has limited sources of income and there was concern that further large oil leaks could have seriously damaged the island's tourism and fishing industries. Consequently, Salvage and Marine Operations were alerted to the problem by the UK Foreign and Commonwealth Office.

Ship plans showed that the RFA *Darkdale* was a single hulled tanker with nine cargo tanks that were longitudinally divided into port, centre, and starboard tanks (creating a total of 27 smaller tanks) and a forward deep cargo tank. The tanker's total oil capacity was around 14,000 tons. Archival research revealed that the RFA *Darkdale* had been refuelled by a Norwegian tanker a few days before it was torpedoed, so it was highly likely that it sank with nearly full tanks.[1] Oil spill modelling indicated that an acute oil spill would impact local fishing and tourism industries so it was concluded that the risk must be mitigated (Liddell & Skelhorn, 2012). In early 2012 a team deployed to St Helena to assess the wreck's condition and its impact on the environment, and to prepare for an intervention.

A side-scan sonar survey confirmed that the wreck is broken in two sections; the bow section is completely inverted, and the stern section lies on its port side (Liddell & Skelhorn, 2012). An ROV survey inspected the condition of the hull and identified five cargo tanks that appeared intact and so could still contain oil. Since each tank is longitudinally divided into three smaller tanks there were 15 potentially intact tanks that needed to be hot tapped. All data gathered were combined with a model of the ship to estimate that 2800–4500 m^3 of oil remained on the wreck.

An initial environmental assessment indicated that the oil was not noticeably elevated in the water column, but it was accumulating in local sediments and occasionally exceeded European Quality Standards. Analysis of local fish and shellfish

[1] Logbook of Norwegian oiler M/T Egerø for the period 23 August—30 October 1941—details provided in a letter from the Riksarkivet—The National Archives of Norway dated 20 October 2011.

generally showed low levels of hydrocarbon contamination, but around 10% of fish sampled exceeded European Quality Standards indicating they could be hazardous to human health if consumed. A wider in-depth survey showed that hydrocarbon concentrations in fish were below the EU regulation safe limit (Cefas, 2014), but a no fishing zone was implemented around the wreck as a precaution.

In 2015 a team deployed to St Helena to remove the remaining oil. The island of St Helena lies around 7000 km south of the UK and there was no functioning airport on the island at the time making this logistically challenging. All equipment and personnel travelled by sea from South Africa, and three ships were brought in from various locations to support the operation. However, the wreck lies just within surface supply diving depth and the bow lying inverted made accessing the tanks straightforward. Initially the plan was to hot tap using a tool designed specifically for wrecks. This tool was lowered onto the wreck using a crane and positioned by surface supplied divers before suctioning on and drilling through the hull, but it turned out to be unsuitable for this wreck. Instead, divers used a manual hot tapping system then attached hoses to pump out the tank contents, all whilst being monitored by the ROV. Most of the tanks held Admiralty Fuel Oil but there was one tank holding aviation gas, and this had deteriorated over the years, presumably due to microbial breakdown (biodegradation), resulting in an acidic product. The operation was successfully completed with 21 tanks hot tapped and approximately 2000 m^3 of oil removed, significantly reducing the risk of a major oil spill from the wreck.

6.5.2 *RFA* War Mehtar

The tanker RFA *War Mehtar* was in convoy FS50 travelling from Methil to Harwich carrying 7000 tonnes of Admiralty fuel oil on the night of 19 November 1941 when it was hit by a torpedo from the German S-Boat S-104 on the portside. The crew abandoned ship and the tug *Superman* took *War Mehtar* under tow. A second tug arrived in the morning, but *War Mehtar* sank and now lies 15 nm east of Great Yarmouth at a depth of 45 m.

The captain remarked in his casualty report that fuel leaked into the engine and boiler rooms after the torpedo attack, but he did not believe it escaped into the sea. However, the rescued engine room crew were 'black with oil', suggesting oil was in the sea. There are also conflicting statements on the extent of the fire. Due to the uncertainty over oil loss the risk was cautiously assessed at full capacity of 7000 tonnes of oil. The potential high volume combined with proximity to shore flagged the *War Mehtar* as a high-risk wreck for further investigation. The *War Mehtar* has undergone several phases of investigation which are described in Hill et al. (2022) and associated reports.

Oil spill modelling predicted that in a worst-case scenario of a total release of oil, around half of it would end up on the shoreline. Some oil sheens and occasional slicks have been reported as coming from the *War Mehtar*, yet water and sediment samples collected from around the wreck did not contain unusually high concentrations of hydrocarbons, so pollution from the wreck does not seem to be accumulating in the environment.

An ROV survey in 2017 showed that the wreck was riddled with holes, but rudimentary tracking on the ROV meant the holes could not be attributed to individual tanks. When oil appeared to be leaking from the wreck in 2018 further investigation was required to determine the contents of each tank, which was done the following year using high resolution ROV-mounted multibeam sonar and neutron backscatter. Neutron backscatter is a technique by which neutrons are fired through a tank bulkhead and neutrons that are returned are measured. The number of neutrons returned depends on the density of hydrogen atoms in the tank, which varies between water and hydrocarbon, so it is possible to detect interfaces between water and oil.

A 3D model of the wreck was created from the multibeam sonar data and compared to data collected during a survey in 2014 to identify any changes to the wreck in the previous four years (WAVES, 2019; Lawrence et al., Chap. 11, this volume; Fig. 6.1). The ship's plans were laid over the 3D model to locate the extent of each of its seven longitudinally divided (port and starboard) oil tanks, as well as its bunker fuel tanks (Fig. 6.2). This meant that each tank could be assessed from the multibeam data and the ROV could be positioned onto each for an individual assessment with the neuron backscatter unit. The multibeam survey showed that nine of the 14 cargo oil tanks had holes and/or cracks and were therefore open to seawater, and the oil fuel bunker tanks had completely collapsed.

The neutron backscatter unit was calibrated against a breached tank before taking readings from apparently intact tanks. All readings indicated water, not oil, within each tank (WAVES, 2019). This evidence gives confidence that, although small volumes of oil are occasionally released from the wreck, there remains no large volume of oil that poses a significant environmental risk.

6.5.3   SS Derbent

The steam tanker SS *Derbent* was on a voyage from Liverpool to Queenstown (the former name of Cobh, Ireland, prior to the 1930s) with a cargo of 3700 tons of fuel oil on 29 November 1917 when it was torpedoed on the port side by U-96. The wreck lies 6 nm north-east of Anglesey, Wales, at a depth of around 45 m. Today, the wreck is regularly dived and in 2014 researchers at Bangor University completed a multibeam sonar survey of the wreck. It is lying on its starboard side facing southeast and, as of 2014, appeared to be intact with the oil tank hatches sealed.

The potential for the *Derbent* to contain as much as 3700 tons of oil combined with its proximity to the shoreline escalated the wreck as a priority for further investigation. There had been no documented oil leaks from the wreck, and oil spill modelling predicted that a major release of oil could impact the British coast and have a high impact on local marine life.

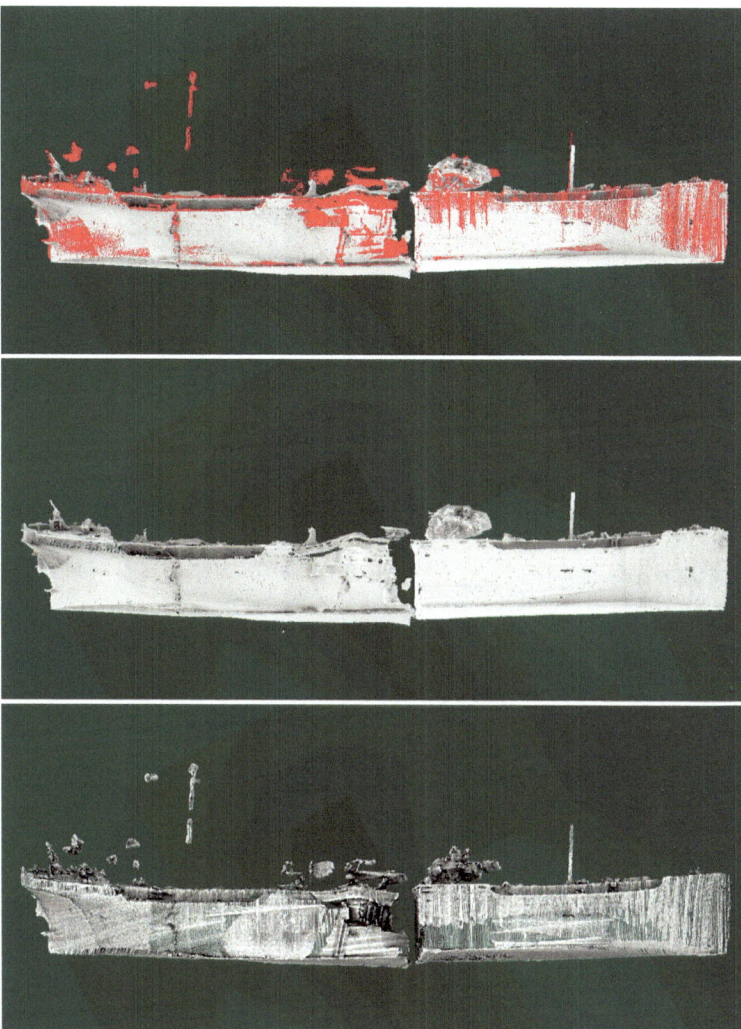

Fig. 6.1 3D model created from multibeam sonar data collected in 2014 (top) and 2019 (middle) and the two combined (bottom: 2019 data in grey and 2014 data in red) to identify changes in the wreck between the 2 years. (WAVES, 2019)

In 2022 a high-resolution ROV-mounted multibeam sonar and neutron backscatter survey was completed to determine the contents of each of the wreck's tanks, as was done on the RFA *War Mehtar* in 2019. The neutron backscatter survey indicated that each tank contained only seawater (WAVES, 2022). The multibeam data collected were compared to the 2014 Bangor University survey and showed almost total collapse of the wreck to one side (see Lawrence et al., Chap. 11, this volume). Despite the collapse, there were no reports of oil washing ashore in the region between the two surveys, suggesting the oil was released decades ago.

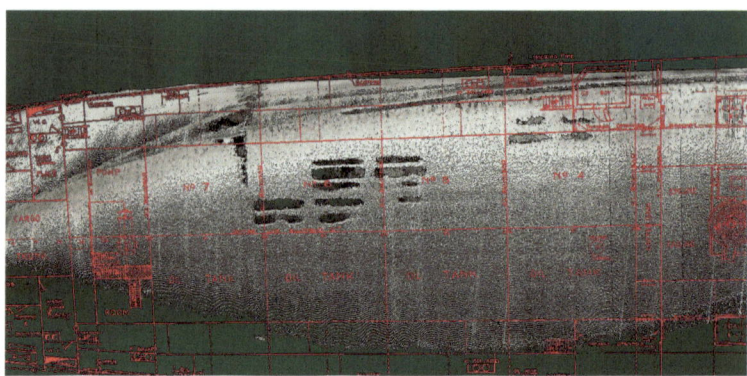

Fig. 6.2 3D model with general assembly plan overlaid, showing the bottom of No.6, No.5 and No.4 tanks with missing plating. (WAVES, 2019)

The significant collapse of the wreck demonstrates how potentially unstable wrecks are when lying on their sides, compared to upright or inverted. This has prompted further analysis of wrecks hitherto surveyed to look for patterns of ship attitude and level of deterioration.

6.5.4 *HMS* Prince of Wales *and HMS* Repulse

The battleship HMS *Prince of Wales* and battlecruiser HMS *Repulse* were sunk by Japanese aircraft on 10 December 1941 off the east coast of Malaysia. The wrecks were both designated as 'Protected Places' under the Protection of Military Remains Act 1986 (PMRA 1986) in 2001. However, this has proved difficult to enforce and, in fact, in international waters PMRA 1986 is only enforceable against British citizens or British-flagged ships interfering with the wreck. Consequently, the wrecks have been salvaged, along with other wrecks in the region including the Japanese Usukan wrecks (Holmes, 9 February 2017a), and HMAS *Perth* and HNLMS *Kortenaer* (Holmes, 3 November 2017b). Although not known for certain, it is generally assumed that the wrecks were targeted because they were built with 'prenuclear' or 'low-background' steel which is needed for the manufacture of Geiger counters and medical equipment (Manders, 2020).

There was naturally a public outcry at the damage to the last resting place of hundreds of crew: 327 on *Prince of Wales* and 508 on *Repulse*. There were also concerns of the impacts on the environment since the *Prince of Wales* and *Repulse* were believed to have each been holding 1000–3300 tons of oil, and the smash and grab techniques of the salvors took little account of the risk that this oil could be released. Oil spill modelling of acute oil spills from the wrecks predicted that south-east Malaysia could be impacted in the northeast monsoon period and southwest

Vietnam could become oiled in the southwest monsoon period. Satellite imagery captured oil leaking from the wrecks in 2014, but no reports of oil washing ashore were apparent.

A team deployed to determine the state of the wrecks and the likely remaining volume of oil in each. If there remained significant quantities of oil it would have been prudent to remove the remaining oil before the rest was released during salvage operations. Speaking with local wreck divers proved invaluable. Not only were they documenting the demise of the wrecks, but they knew their exact location which was most useful when the official coordinates turned out to be inaccurate.

The wrecks were surveyed using towed side-scan sonar and high resolution ROV-mounted multibeam sonar (Salvage and Marine Operations, 2019; Fig. 6.3). The imagery showed extensive damage to both wrecks, particularly *Repulse*, perhaps since it lies in slightly shallower water (54 m compared to 68 m in the case of *Prince of Wales*). In addition to causing oil spills, the physical damage to the wrecks resulting from the use of explosives and grabs has severely impacted the marine ecosystem that had developed around the wrecks. Photographs taken by recreational divers in the years prior to the salvage show that *Prince of Wales* and *Repulse* were once home to a diverse community of marine life, including corals and sponges, providing havens for small fish and food for parrotfish, manta rays and sea turtles. The explosives used to gain entry to the wrecks have dislodged the organisms that once encrusted them and removed the habitat. The hard coral now lies in bone-like fragments on the seabed; the wrecks are colourless and lifeless except for a few fish.

When the sonar images were cross-referenced with ships plans it was possible to identify which tanks could remain intact. Based on this, it was estimated that a maximum of 400 m³ and 250 m³ remained on *Prince of Wales* and *Repulse*,

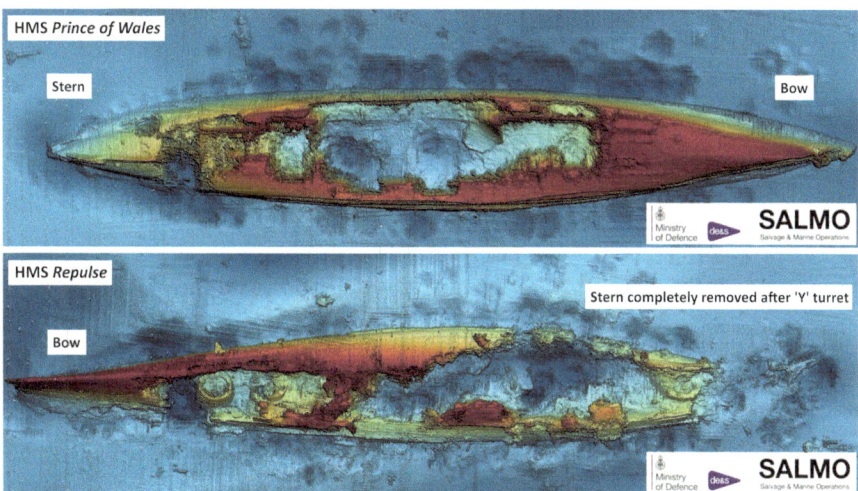

Fig. 6.3 Multibeam sonar image of HMS *Prince of Wales* (top) and HMS *Repulse* (bottom). (Salvage and Marine Operations, 2019)

respectively. This is a significant volume of oil, but there was evidence that more salvage had been occurring prior to our arrival, suggesting it was an ongoing process. It therefore seemed unlikely that significant quantities of oil would remain on the wrecks by the time an intervention could be planned and initiated.

6.6   Summary

Although wrecks have always been dealt with reactively in the UK, it is only since the 1990s that their environmental impacts have been considered, and since the inception of the Wreck Management Programme in 2008 that they have been actively managed. This relatively recent change to the UK's approach to wreck management is in response to changing sensibilities and understanding around the environmental and human health impacts of oil pollution in the sea.

Wreck management in the UK evolves as more is learnt about the subject area, including identifying gaps in knowledge. Assessing PPW in UK waters keeps the UK government busy but there is also a great feeling of responsibility to those living with British PPW in their waters overseas and an awareness that some people will be more impacted by a leaking wreck than others (Hill, 2021). The UK government does not adopt an out of sight, out of mind policy, and instead actively engages with the governments of nations where the UK has left behind legacies of war.

Despite attempts to proactively manage the risks associated with British PPW, the sheer quantity (around 10,000 wrecks globally) compared to available resources means that the proactive side falls far short of sufficient, and consequently much work is done in reactive mode. However, as demonstrated through the case studies, there is an ongoing programme of work that aims to mitigate risks before they materialise.

References

Bergstrøm, R. (2014). Lessons learned from offloading oil from potentially polluting ship wrecks from World War II in Norwegian waters. *International Oil Spill Conference Proceedings, 1,* 804–813. https://doi.org/10.7901/2169-3358-2014.1.804

Brennan, M. L., Delgado, J. P., Jozsef, A., Marx, D. E., & Bierwagen, M. (2023). Site formation processes and pollution risk mitigation of World War II oil tanker shipwrecks: *Coimbra* and *Munger T. Ball. Journal of Maritime Archaeology, 18,* 321–335.

Cefas. (2014). St Helena edible fish contamination study, Final project report (C6130), 117 pp.

Goodsir, F., Lonsdale, J. A., Mitchell, P. J., Suehring, R., Farcas, A., Whomersley, P., Brant, J. L., Clarke, C., Kirby, M. F., Skelhorn, M., & Hill, P. G. (2019). A standardised approach to the environmental risk assessment of potentially polluting wrecks. *Marine Pollution Bulletin, 142,* 290–302.

Hill, P.G. (2021). *Is the Ministry of Defence wreck environmental desk-based assessment really globally applicable? A case study of the wreck of the USS Mississinewa in Ulithi Lagoon, Federated States of Micronesia.* MSc Dissertation, University of Bristol, 41 pp.

Hill, P. G., Skelhorn, M., & Leather, S. (2022). Assessing the environmental risk posed by a legacy tanker wreck: A case study of the RFA War Mehtar. *Environmental Research Communications, 4*, 055005.

Holmes, O. (2017a, February 9). Images reveal three more Japanese WWII shipwrecks torn apart for scrap. *The Guardian.* https://www.theguardian.com/world/2017/feb/09/images-reveal-three-more-japanese-wwii-shipwrecks-torn-apart-for-scrap

Holmes, O. (2017b, November 3). The world's biggest grave robbery: Asia's disappearing WWII shipwrecks. *The Guardian.* https://www.theguardian.com/world/ng-interactive/2017/nov/03/worlds-biggest-grave-robbery-asias-disappearing-ww2-shipwrecks

Kelly, C., Law, R. Emerson, H. (2000). Methods for analysis for hydrocarbons and polycyclic aromatic hydrocarbons (PAH) in marine samples: *Science Series, Aquatic Environment Protection: Analytical Methods* 12, 18 pp.

Liddell, A., & Skelhorn, M. (2012). RFA Darkdale Survey Report, Salvage and Marine Operations, 173 pp.

Liddell, A., & Skelhorn, M. (2018). Deriving archaeological information from potentially-polluting wrecks. In K. Bell (Ed.), *Bridging the gap in maritime archaeology: Working with professional and public communities* (pp. 81–89). Archaeopress Publishing Ltd..

Manders, M. R. (2020). The issues with large metal wrecks from the 20th century. In A. Hafner, H. Öniz, L. Semaan, & C. J. Underwood (Eds.), *Heritage under water at risk: Challenges, threats and solutions* (pp. 73–77). The International Council on Monuments and Sites.

Provision of Mutual Aid Concerning Certain Problems of Marine Transportation and Litigation Agreement, United Kingdom-United States, December 4, 1942, Treaty Series No. 1. (1943). https://digital-commons.usnwc.edu/cgi/viewcontent.cgi?article=2221&context=ils

Salvage and Marine Operations. (2019). Stage 2 wreck assessment report for HMS PRINCE OF WALES and HMS REPULSE, 42 pp.

Thomas, G. E., Bolam, S. G., Brant, J. L., Brash, R., Goodsir, F., Hynes, C., McGenity, T. J., McIlwaine, P. S. O., & McKew, B. A. (2021). Evaluation of polycyclic aromatic hydrocarbon pollution from the HMS Royal Oak shipwreck and effects on sediment microbial community structure. *Frontiers in Marine Science, Marine Pollution, 8*, 650139.

Viduka, A. (2020). Public engagement, community archaeology, underwater cultural heritage management and the impact of GIRT scientific divers in New Zealand. *Australasian Journal of Maritime Archaeology, 44*, 21–38.

WAVES. (2019). RFA War Mehtar wreck assessment report. Report for the Ministry of Defence, ref.:CW1252-R02.SWL, 29 pp.

WAVES. (2022). Survey and assessment of remaining hydrocarbons on the wreck of the SS DERBENT Report for the Ministry of Defence, ref.: CW1610.R01.SL, 33 pp.

Open Access This chapter is licensed under the terms of the Creative Commons Attribution 4.0 International License (http://creativecommons.org/licenses/by/4.0/), which permits use, sharing, adaptation, distribution and reproduction in any medium or format, as long as you give appropriate credit to the original author(s) and the source, provide a link to the Creative Commons license and indicate if changes were made.

The images or other third party material in this chapter are included in the chapter's Creative Commons license, unless indicated otherwise in a credit line to the material. If material is not included in the chapter's Creative Commons license and your intended use is not permitted by statutory regulation or exceeds the permitted use, you will need to obtain permission directly from the copyright holder.

Chapter 7
Polluting Wrecks in the Baltic Sea

Benedykt Hac

7.1   Introduction

The Baltic Sea is a shallow inland sea on the continental shelf of northern Europe. The shores and bottoms in the area of Sweden, Finland, Russia and Estonia are mainly rocky with skerries, while the shores and bottoms in the area of Latvia, Lithuania, the enclave of Königsberg (Russia), Poland, Germany and Denmark are mostly flat and sandy with sand, clay and silt bottoms. Water exchange with the North Sea and the Atlantic Ocean is mainly through the shallow Danish Straits, Kattegat and Skagerak, resulting in a salinity range of 2–12‰ due to intermixing from riverine input. The area of the Baltic Sea is 392,979 km^2, the water volume is 21,721 km^3 and the average depth is 52.3 m with a maximum depth of 459 m (Baltic Sea, 2023).

Tidal fluctuations average between 3 and 24 cm. There are seiches (max. 30 cm) and water level changes due to wind (max. 200 cm). The sea is often turbulent and can create waves that are short and steep. The maximum recorded wave height was 14 m (24.12.2004) with a wave length of 50 m. The Baltic Sea is a particularly sensitive area to pollution from fuels, oils, and chemicals such as fertilisers, sewage and substances from dumped chemical and conventional munitions. These inputs into the marine system are compounded due to the very low exchange of the Baltic Sea with the Atlantic Ocean because of the narrow outflows, and the residence time can be 25–30 years; this results in a low capacity of the basin to self-purify and flush pollutants and contaminants from the system. Therefore, pollution from sunken ships in the Baltic Sea poses a significant threat to the ecosystem.

B. Hac (✉)
MEWO S.A., Straszyn, Poland

© The Author(s) 2024   85
M. L. Brennan (ed.), *Threats to Our Ocean Heritage: Potentially Polluting Wrecks*, SpringerBriefs in Underwater Archaeology,
https://doi.org/10.1007/978-3-031-57960-8_7

7.2 History of Operations

As of 2005, there were an estimated 8569 potentially polluting shipwrecks in the world's oceans, of which 1583 were oil tankers. Approximately 75 percent of these wrecks sank during the Second World War. More than 70 years later, these wrecks are a major source of potential pollution and represent a global marine pollution risk (Michel et al., 2005). The wrecks on the bottom of the Baltic Sea, containing heavy and light fuel oil, represent merchant and naval vessels from the period of both World Wars, as well as vessels sunk in the twentieth and twenty-first centuries as a result of collisions or storms that have occurred since the wars.

The Gulf of Finland, the Gulf of Riga and its approaches, the Gulf of Gdansk and the southern Baltic Sea, as well as narrow areas such as the Danish Straits and the approaches to the Kiel Canal, saw particularly intense warfare in which many ships and other liquid-fueled vessels were sunk. In just one operation, the evacuation of troops and civilians during Operation Hannibal by the German Kriegsmarine in 1945, some 250 vessels were sunk, many of them carrying large quantities of fuel. In total, it is estimated that there are around 1000 wrecks in the Baltic Sea alone, some with unspecified, sometimes significant quantities of fuel (for their own use as well as cargo) in their holds or tanks.

Scientific institutions (IMUMG—Maritime Institute of Maritime University of Gdynia, IOPAS—Institute of Oceanology Polish Academy of Science, SYKE—Finnish Environment Institute) and the maritime administrations of the Baltic States are still investigating what percentage of the wrecks are dangerous for the environment. Given the passage of time and the progressive corrosion of the wrecks (steel loss of about 0.01 to 0.1 mm/year), the risk of major spills and pollution of the waters and beaches of the Baltic Sea is rapidly increasing (Pärt et al., 2015). Time, anthropogenic activities in the water such as fishing, stormy weather in shallow areas and hull degradation due to corrosion, all significantly increase the possibility of large spills. Currently, large spills from wrecks are occasionally recorded, but small spills or leaks from these wrecks are common.

Periodically, large slicks of fuel appear in the open sea, the origin of which is difficult or impossible to determine. Very often these are likely leaks from wrecks. The use of VTS (Vessel Traffic Services) in busy areas and AIS (Automatic Identification System) throughout the Baltic Sea has meant that uncontrolled discharges of fuel from active vessels have been stopped and, if they do occur, the offender can be found within a matter of hours. Due to the small area of the Baltic Sea, each piece of its seabed has a clearly defined ownership. The areas of responsibility of the individual countries overlap and cover the entire area of the sea.

7.3 Wrecks in the Baltic Region

Sweden, Finland, Russia, Estonia, Lithuania, Poland, Germany and Denmark maintain their own services to identify potential pollution hazards. The knowledge of the subject, the capabilities and modus operandi, and the institutions monitoring the situation vary greatly and are at very different levels of organisation and technical sophistication.

7.4 Sweden

The Swedish National Maritime Museums maintains a catalogue of data on 17,000 underwater objects in its waters. This catalogue has now been integrated into the FMIS (Informationssystemet över fornminnen) and the Historic Sites Information System. It has been estimated that 316 wrecks may have some fuel resources, but only 31 wrecks pose a significant threat to the environment (Sjöfartsverket, 2014). These are Cat 1 wrecks containing more than 100 tonnes of fuel. At the same time, there are ongoing efforts in Sweden to clean up the wrecks that have been identified as hazardous.

To this end, Chalmers University has developed the VRAKA assessment system, and the Swedish Maritime and Water Administration effectively cleans up 2 or 3 wrecks per year. To do this, it has a fixed annual budget of 2.4 million euros allocated in advance for 10 years. This makes it possible to establish a sustainable system for planning, monitoring and removing dangerous wrecks.

A good example of this is the wreck of the ship *Skytteren*, which sank on 1 April 1942 about 10 km off the west coast of Sweden near the town of Lysekil in the Kattegat (VRAK Museum of wrecks, 2023). During an attempt to break a naval blockade, the ship was damaged by the Germans. To prevent the Germans from seizing the valuable cargo of Swedish steel and ball bearings, the captain decided to sink the ship. According to the environmental risk assessment, *Skytteren* is number one on the list of the 30 most environmentally hazardous wrecks. It is estimated that the wreck may contain up to 400 cubic metres of oil. The location of the wreck is already outside the Baltic Sea proper, but in the main strait leading into the Baltic, which means that it is an excellent example of coordinated action by the Swedish services as a case study for oil clean-up operations at greater depths (74 m).

According to the Swedish website Vrak Museum of Wrecks, 'In the winter of 2005, large quantities of oil appeared in the waters north of Måseskär on the west coast of Sweden. The oil came from *Skytteren* and at its peak the discharge was almost 400 litres per day.' The *Skytteren* spill was a wake-up call to a problem that had previously been overlooked.

By 2018, work had begun to remove environmentally hazardous substances from *Skytteren*. After extensive investigations in 2021 and 2022, 175,000 L were recovered by drilling holes in the hull and pumping out the oil and oil-water mixture. However, the wreck may still contain oil because *Skytteren* is on its side and some of the fuel tanks are inaccessible.

7.5   Finland

Finland has an interactive database, produced by SYKE, the Finnish Environmental Institute, showing the distribution of known wrecks (Finnish Wreck Base, 2023), with 5200 records, of which about 1000 are wrecks, of which 420 wrecks contain fuel, and of those, 12 are Cat. 1 wrecks (more than 100 tonnes with certainty), 24 are Cat. 2 wrecks that may contain more than 100 tonnes (with less certainty), 68 are Cat. 3 wrecks expected to contain between 10 and 100 tonnes of fuel and could be considered potentially hazardous, and the remainder contain less than 10 tonnes (Rytkönen, 1999, Jolma, 2009). As of 2019, the concept for the operation of the base has changed. The search, survey and clean-up of the wrecks is carried out by the Ministry of the Environment, the Finnish Coast Guard and the Finnish Navy in cooperation with SYKE. These institutions have been continuously provided with financial and material resources (ship *Aranda*) to carry out wreck exploration and clearance operations.

7.6   Russia

There are large amounts of wreckage, including potentially environmentally hazardous wrecks, in both the Gulf of Finland and the Königsberg enclave areas of western Russia. This is a result of the wars that took place in the area of both locations. Based on historical literature, it can be estimated that there are at least several hundred such wrecks. Unfortunately, there is no access to the official wreck database of the Russian Hydrographic Office. The only visible fact that such knowledge exists in Russia are the positions of known wrecks marked on nautical charts (both Russian, Finnish, Polish and English). These charts do not define the size or nature of the wrecks, other than that they are dangerous to navigation. It is therefore difficult to estimate how many of them are potentially hazardous to the marine environment. Preliminary data collected by J. Rytkönen (2021) for the study for the HELCOM Submerged project indicate a high potential for oil pollution in the eastern Gulf of Finland, which is in Russian territorial waters. Private divers report that the following wrecks lie in the Russian part of the Gulf of Finland: 7 destroyers (300–500 tonnes of oil), several minesweepers and large submarine chasers (up to 100 tonnes of fuel), 8 submarines (60–120 tonnes of fuel), 4 minesweepers and gunboats (30–100 tonnes of fuel), 3 large merchant ships (with possibly hundreds of tonnes of fuel), and a bunker tanker (1000 tonnes of petrol). It is thought that the 35 known wrecks may contain some 843 tonnes of light fuel oil and 5870 tonnes of heavy fuel oil. How much of this fuel remains in the tanks 80 years after sinking is entirely unknown.

7.7   Estonia

As a result of fierce naval battles and difficult navigational conditions, there are many wrecks in Estonian waters. According to Estonian sources, about 705 wrecks have been located (SHIPWHER, 2010), of which 54 have been identified as potentially polluting, and 13 of these have been investigated in detail. Wrecks are registered in the central wreck register SHIPWHER, which is available online. The Estonian Maritime Administration is responsible for investigating wrecks and must remove those that pose a safety or environmental risk. The Estonian Maritime Administration is also responsible for mapping and registering wrecks along the waterways (Svensson, 2010). The database currently contains 1297 objects on the seabed. In July 2023, it was reported that the wreck of HMS *Kasandra*, which sank west of the large Estonian island of Saaremaa after the First World War, had been found (The Baltic Times, 2023). The Climate Ministry reported that, 'The ship is thought to have carried around 780 tonnes of fuel, some of which was probably in the ship the time of the sinking. Given the relative integrity and condition of the wreck, it is believed that most of the fuel remains trapped within the wreck.'

7.8   Latvia

There is no access to an official shipwreck database of Latvia, which is presumably held by the Latvian Maritime Administration (Svensson, 2010), which ensures safe navigation in its waters. Experts from the Ministry of the Environment reported in 2010 that there had been cases of assessment and removal of shipwreck contamination, but these concerned recently sunken ships in port areas. No visible contamination was found in other areas. The unofficial register of shipwrecks in Latvia, maintained by the private association of divers 'Nirēju klubs Daivings', lists about 200 wrecks in Latvian waters. There are at least a few wrecks from the First World War, which gives them the status of archaeological heritage. Most of them are steamships, but some of them pose a certain environmental risk due to the armaments and shells on board. An unspecified number of ships and vessels sunk during World War II contain wrecks that may pose an environmental risk.

7.9   Lithuania

There is no official information on the number of wrecks in Lithuanian waters. Recently it was reported that there are only seven, which is fewer than expected due to the historical conditions and war losses in this area during World War II. Interestingly, on the Polish map 'Wraki Bałtyku', a private Polish company that organises wreck diving, there are ten locations of large wrecks of interest to divers that lie close to the Lithuanian coast. The Maritime Safety Administration is the only institution capable of conducting wreck surveys, as it is the only one with hydrographic survey vessels equipped to conduct underwater surveys (Svensson, 2010). Recent small spills recorded on the surface suggest that there are several wrecks on the seabed that contain some fuel and are leaking. There is no information on ongoing clean-up operations or monitoring of dangerous wrecks.

7.10   Poland

The official database of wrecks in Poland is maintained by the Hydrographic Office of the Polish Navy (Database of Underwater Objects—BDoP) and contains several thousand records (in 2023—total 10,502), including about 530 wrecks (June 2023). Experts believe that there are more, in the range of 600 to 800, many of which have not yet been found or identified. A project to assess the environmental conditions of the seabed in parts of the Polish Exclusive Economic Zone (EEZ) is currently underway on behalf of the Chief Inspectorate for Environmental Protection (GIOS); at this stage, surveys have been carried out in up to 30 percent of the EEZ and will continue up to about 70 percent of the EEZ. Several dozen new wrecks are discovered each year (in 2023–43 wrecks and 808 objects). Previously, wrecks and information were collected by the Maritime Institute in Gdansk (IMG) and the Institute of Oceanology of the Polish Academy of Sciences (IOPAS in Sopot). After the takeover of the Maritime Institute in Gdansk by the Maritime University in Gdynia, IOPAS took over the management of the wreck projects, but their interest has become more focused on biological research.

According to IMG's earlier assessment, about 18 of the wrecks investigated are classified as Cat. 1 and 2 ($\geq$100 tonnes of fuel, of which 2 are ultra-hazardous, i.e., *Stuttgart* and *Franken*), and about 50 are Cat. 3 wrecks (10–100 tonnes). The remainder are likely to be Cat. 0, i.e., not hazardous to the environment. At present, the maritime administration has no planned measures to prevent pollution, no wrecks have been cleaned up to date and ongoing surface pollution is dealt with by SAR (Search and Rescue). Some of the more dangerous wrecks have been given special protection due to the high number of casualties from the Second World War. There is ample evidence that these wrecks may have large quantities of oil in their tanks. However, due to the sensitivity of the work on the so-called wet graves (e.g., *Wilhelm Gustloff* 10,000 victims, *Goya* about 5000 victims, *Steuben* about 6000 victims), the start of detailed investigations has been postponed. There are currently no plans for hazard identification on these wrecks.

7.11   Germany

The official database of underwater objects, including wrecks in German waters, is maintained by the Federal Maritime and Hydrographic Agency (BSH) and contains approximately 2500 objects (Svensson, 2010). This information is mainly used for maritime safety (Bundesamt, 2023). Objects such as large rocks, containers and others are also included in the database. The Wikipedia wreck list currently lists 24 wrecks in German waters, most of which are warships and submarines from the Second World War. Although most of the wrecks are on the North Sea side, there are also wrecks near Kiel, Trawemünde and Warnemünde. The wreck map published by the BSH suggests that there are many more wrecks. This is also suggested by the nautical charts published by the BSH, which show the position of dozens of wrecks in the western Baltic Sea. There is a lack of information on current activities to review the status of wrecks and their environmental impact and potential danger. In Germany, which is a federal state, the coastal state is responsible for assessing and dealing with wrecks. In the Western Baltic these are Mecklenburg-Western Pomerania and Schleswig-Holstein.

7.12   Denmark

The Danish Maritime Safety Authority (DMSA) maintains a database of known wrecks in Danish waters. The purpose of the database is to maintain the safety of navigation. There are 2518 known wrecks in the register. For 90% of the wrecks registered before 2000, the data contains only information on the safe depth above the wreck, the depth around the wreck and the position. About a third of the wrecks described in the database are located in the waters of the Baltic Sea and the Danish Straits. The wreck register data is stored in an object class in the ESRI ArcGIS geodatabase. The data and documentation are in Danish. One major wreck removal operation is known, which is of the wreck of the *Fu Shan Hai*, sunk in 2003. In total, approximately 1500 tonnes of heavy fuel oil, light fuel oil and 35 tonnes of lubricating oil were recovered from the wreck or collected from the surface. In 2013, 620 m^3 of water/oil mixture and 251 m^3 of fuel were recovered from the wreck during a 50-day hot tapping clean-up operation. A detailed description of the salvage operations can be found in (Danish Navy, 2013) and (Mortensen & Rasmussen, 2013).

7.13   Wreck Risk Assessment Methods

There are several centres in the Baltic Sea region that use different methods to assess the risk of marine pollution from potentially hazardous wrecks. The most commonly used method is the Swedish VRAKA, recognised by the Swedish,

Finnish and Polish representatives in Helcom Submerged. Other EU-related methods include the DEEPP project risk indicator method developed in the EU-funded SWERA project and the NOAA RULET risk methods. The HELCOM Submerged work proposed a risk approach similar to that used in the US. Poland has also developed its own for English E-DBA risk model protocol, which appears to be the most appropriate to adapt for the assessment of Polish wrecks. This method has a clear and simple structure, but does not take into account the change in risk level over time. However, it provides an unbiased risk assessment and identification of the most appropriate management strategy, thus minimising conflicts of interest with the Maritime Administration, which would be consulted throughout the assessment process (Hac & Sarna, 2021). The Finnish method is characterised by moving away from 'risk assessment' and focusing on factors that determine how easy or difficult it may be in a given case to remove fuel tanks from a wreck. When assessing individual centres, it is important that operations are optimised for the local conditions in which they operate (Table 7.1).

Table 7.1 Summary of shipwreck clean-up operations carried out in the Baltic Sea area

Coastal state	Vessel	Year of sinking	Oil on wreck	Oil removed	Cost
Finland	*Park victory*	1947	600 m³	410 m³	8600€/m³
	Estonia	1994	418 m³	250 m³	
	Brita Dan	1964		20 m³	
	Fuel barge	1918		700 m³	
	Hanna Marjut	1985		20 tonnes comb.	
	Fortuna	1987		20 tonnes comb.	
	Simon		20 m³	0	
	Beatrice		20 m³	0	
Sweden	*Thetis*	1985		730 L	€512,000
	Sandön and *Hoheneichen*				€532,000
	Lindesnäs	1957	NA	299 m³	€2,000,000
	Finnbirch	2006	NA	60 m³ diesel 88.2 m³ HFO	€2,160,000
	Skytteren	1942	400 m³	NA	€931,000
Estonia	*Rubber IV*	2006		100 m³	
	Torpedoboat 20	1944	370 tonnes		
Denmark	*Fu Shan Hai*			1535 tonnes	

7.14 Case Studies of the *Stuttgart* and *Franken* Wrecks

7.14.1 SS Stuttgart

The wreck of the hospital ship SS *Stuttgart* lies on the bottom of the Gulf of Gdansk at a depth of 22 m, approximately 4.4 km northeast of the approach to the Port of Gdynia (Fig. 7.1). SS *Stuttgart* was built in 1923 at the Vulcan—Werke Hamburg und Stettin AG shipyard in Szczecin. From 1930, SS *Stuttgart* sailed on the Far Eastern Lines. With war looming, the ship was conscripted into the German Navy and converted into a hospital ship, known as *Lazaretschiff* 'C'.

The ship's fate was sealed in 1943 during the raid on Gdynia, which as a shipyard and naval port became a target for Allied air strikes. The air raid took place on 9 October 1943 when, among other things, the hospital ship *Stuttgart* was bombed, which for unknown reasons had been covered with camouflage netting and was not recognised as a hospital ship. The ship was carrying a large number of wounded soldiers from the Eastern Front. The number of wounded is still not known. Almost all of them perished in the flames. It is estimated that around 700 people died in the attack. The burning *Stuttgart* posed such a serious danger to other ships in the harbour that it was towed to the roadstead in Gdynia and deliberately sunk with the bodies of the victims. This operation was carried out by the guards of the third escort flotilla using 25 88-mm shells.

Fig. 7.1 From left. Lazaretschiffe C (S/S *Stuttgart*) (*German National Archive*), Multibeam sonar map of the wreck of S/S *Stuttgart* (*Hac B., Maritime institute in Gdansk 2016*), Samples of the soil taken in the area of the wreck using a VanVeen sampler and a vibrocorrer (*Hac B., Maritime institute in Gdansk 2016*)

An attempt was made in 1950 to develop a method of raising the wreck from the seabed. Due to a lack of suitable equipment and documentation, the attempt failed. In 1955, the Polish Salvage Commission gave its final verdict on the sunken wreck, based on the testimony of divers who had surveyed the wreck in 1952 and 1953. The Commission concluded that the hull was 80% damaged, while the wear on the machinery, due to its age and 10 years in the water, was estimated at 100%. The wreck of SS *Stuttgart* was considered an obstruction to navigation and removed by demolition using explosives. A pyrotechnic method was used for heavily damaged wrecks and to remove obstructions to navigation. The wreck was ripped apart with explosives and the parts were removed with cranes. Removal of the wreck began in 1957.

It should be noted that there is no report on the demolition of the wreck in terms of environmental impact. An undetermined amount of fuel leaked from the wreck's tanks during the demolition. Surveys in the area of the wreck site in 2009 revealed significant scattering of ship remains on the seabed. This suggests that *Stuttgart* was scuttled, indicated by the hull, superstructures and the part of the hull that was accessible after the removal of any steel structures that managed to separate from the wreck. A multibeam echosounder survey on 9 May 2015 provided a detailed picture of the remains of the wreck on the seabed.

The Maritime Institute in Gdansk conducted a seabed survey around the wreck in an area of approximately 1.5 km^2. The survey was carried out using a multibeam echosounder, towed side-scan sonar and a sub-bottom profiler. A total of 1050 bottom and water samples were taken for chemical, toxicological and biological analysis using a Van Veen corer, and 50 sediment cores up to 3 m in length were taken using a vibracorer. An ROV visual inspection was also conducted.

In the area of the seabed surveyed, contamination of the seabed with synthetic heavy fuel oil from the Fisher-Tropsh process was detected (Hac, 2014). Contamination was found with a thickness ranging from 5 to 130 cm over an area of approximately 415,000 m^2 (Fig. 7.2). The levels of polycyclic hydrocarbons exceed standards by about 3000 times (Rogowska & Wolska, 2019). High levels of many heavy metals exceed the standard locally by up to 18,000 times. Locally, there are large lakes of liquid fuel deposited in the subsurface depressions. We estimate that about 600 to 1000 tonnes of synthetic fuel, which is heavier than water, is deposited on the seabed.

Several ideas were put to the Maritime Administration to solve the problem of soil contamination on the seabed. One was to remove the entire top layer of contaminated soil, clean it of oil and place it back on the seabed. This is not possible due to the size of the one million cubic metres of contaminated soil and regulations that prohibit the placement of contaminated soil on the seabed. The cost of this measure has been estimated at €150 million.

Another approach proposed was to cover the area with a special mineral-based, chemically neutral mixture that would allow the fuel-eating bacteria to develop and grow rapidly, creating favourable conditions for their growth and accelerating natural soil remediation. This is the most environmentally friendly method. The cost of

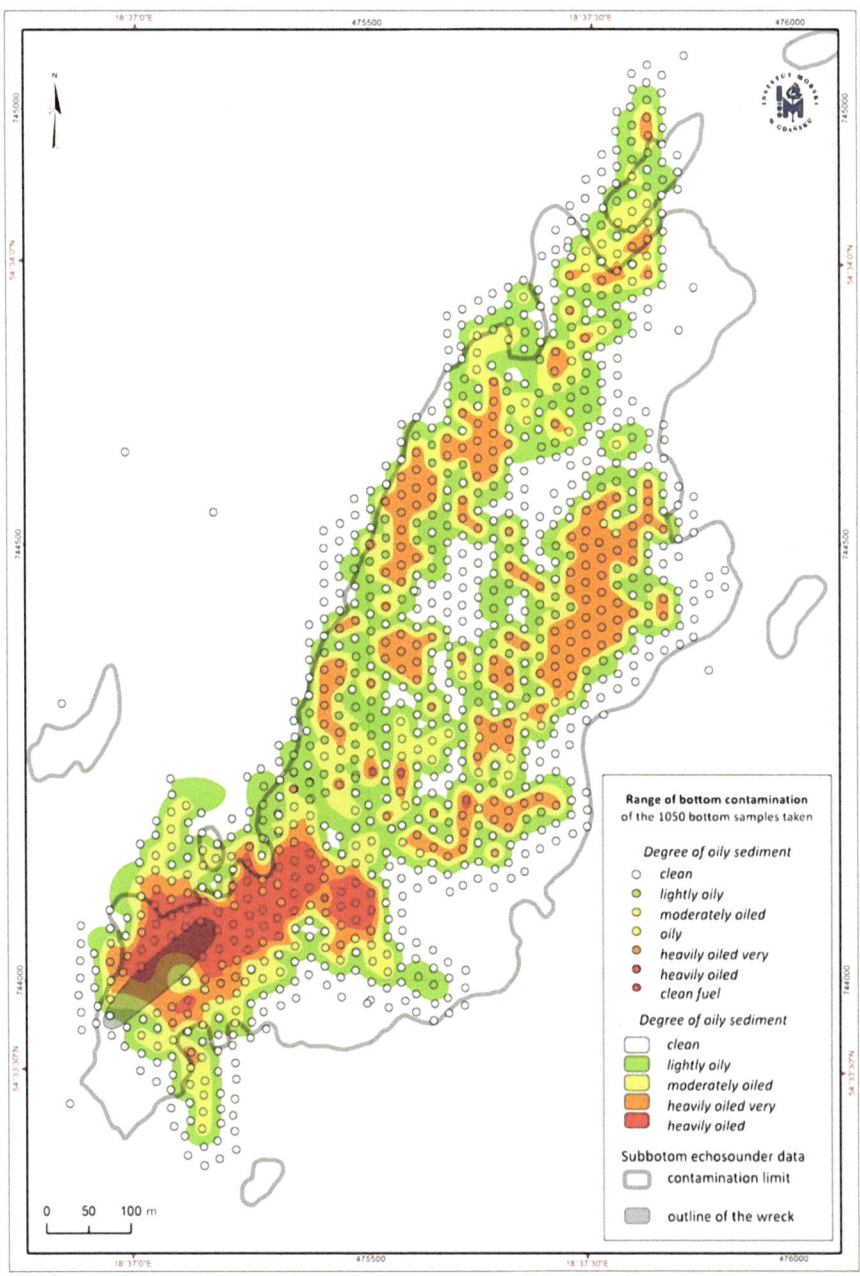

Fig. 7.2 Range of bottom contamination of 1050 bottom samples taken around wreck of S/S *Stuttgart* (*Hac B., Maritime Institute in Gdansk 2016*)

the operation was estimated at around €15 million. Very detailed monitoring of the extent of contamination and changes in the soil over the past 8 years, its natural remediation capacity, is currently underway.

7.15   Supply Vessel *Franken* (Trossschiff *Franken*)

Franken was a 179-metre-long, motor-driven tanker/supply ship with a displacement of 22,850 tonnes, construction of which began in 1937 at the Deutsche Werke shipyard in Kiel (Fig. 7.3). The hull was launched on 8 March 1939. In 1942, it was towed to Burmeister & Wain in Copenhagen, where it was completed and commissioned on 17 March 1943.

Fig. 7.3 From left. Supply Vessel *Franken* (Trossschiff 'Franken') (*German National Archive),* Loading plan of Supply Vessel *Franken* (Trossschiff 'Franken') (*German National Archive)* Multibeam sonar map of the wreck of S/V *Franken* (*Hac B., Maritime institute in Gdansk 2016)*

It was used as a supply ship for the battleship *Prinz Eugen* and the ships of the armoured group Tiele, and also supplied torpedo boats and minesweepers. The ship was sunk on 8 April 1945 by the Soviet air force. The supply vessel was capable of carrying 9500 tonnes of fuel (heavy oil, light oil, aviation gasoline), 306 tonnes of lubricating oils of various types, 973 tonnes of ammunition (calibres from 20 mm to 280 mm), 822 m^3 (790 tonnes) of spare parts, and technical supplies for ships.

The last information on the amount of fuel on board, which was intercepted by Allied intelligence 8 days before the sinking, was 3060 m^3 of fuel. During these 8 days a considerable amount was transferred to other ships and an unspecified amount spilled on the surface when the bombed ship had broken. We estimate that between 600 and 1000 tonnes, mostly heavy fuel oil, remained in the surviving tanks. Investigations have shown that in the trough around the wreck, an area of about 3 hectares, the ground is saturated with a large amount of heavy fuel (Hac, 2018).

7.16 Conclusion

Because of their vulnerability to pollution in the Baltic Sea, the countries bordering the Baltic Sea have taken significant steps to assess the number of hazardous wrecks, including addressing how to assess the risk of oil spills, how to clean them up by removing the deposited fuel and how to rehabilitate the contaminated land around the wrecks. All Baltic countries cooperate in the oil spill clean-up system. Finland and Sweden have a good track record in cleaning up wrecks. Other countries have made progress in developing and implementing systems to locate and remove contamination from wrecks. There are activities under the European South Baltic Project and other projects that target the Baltic Sea as a region of particular environmental concern. A two-pronged effort is currently underway in Poland to clean up known wrecks that pose an environmental risk. On the one hand, through participation in the above-mentioned South Baltic projects, experiments are being carried out to introduce new methods of seabed remediation (the *Stuttgart* wreck) and full information is being obtained on 18 wrecks in the Polish EEZ (including the *Franken* wreck) suspected of containing various types of fuel. On the other hand, state institutions such as the Maritime Office have started to prepare the information needed to set up a fundraising system and to issue calls for tenders for the clean-up of wreck tanks. The first wrecks will be cleaned within the next 5 years. The attached algorithm shows how such an operation could be carried out (Figs. 7.4 and 7.5).

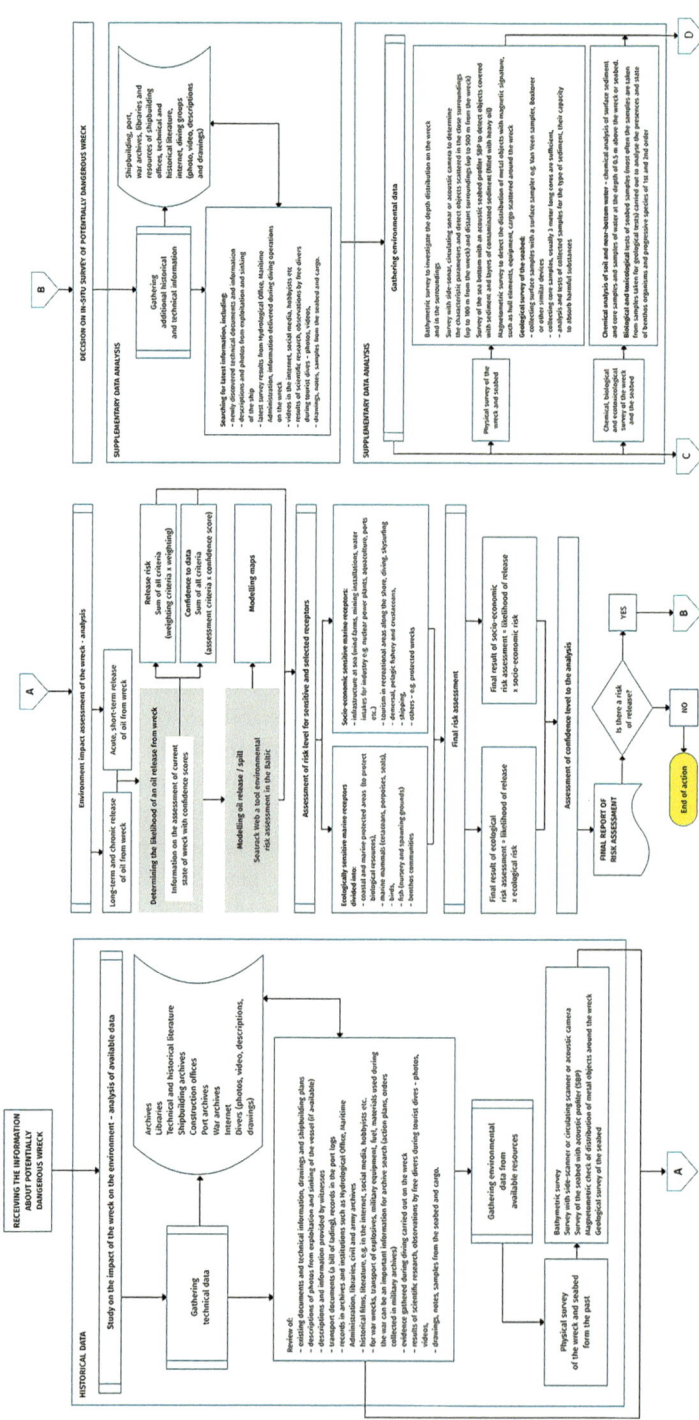

Fig. 7.4 Steps to be taken from the discovery of the wreck, through its identification and risk assessment, to the removal of fuel from the wreck—Part I (Hac & Sarma, 2021)

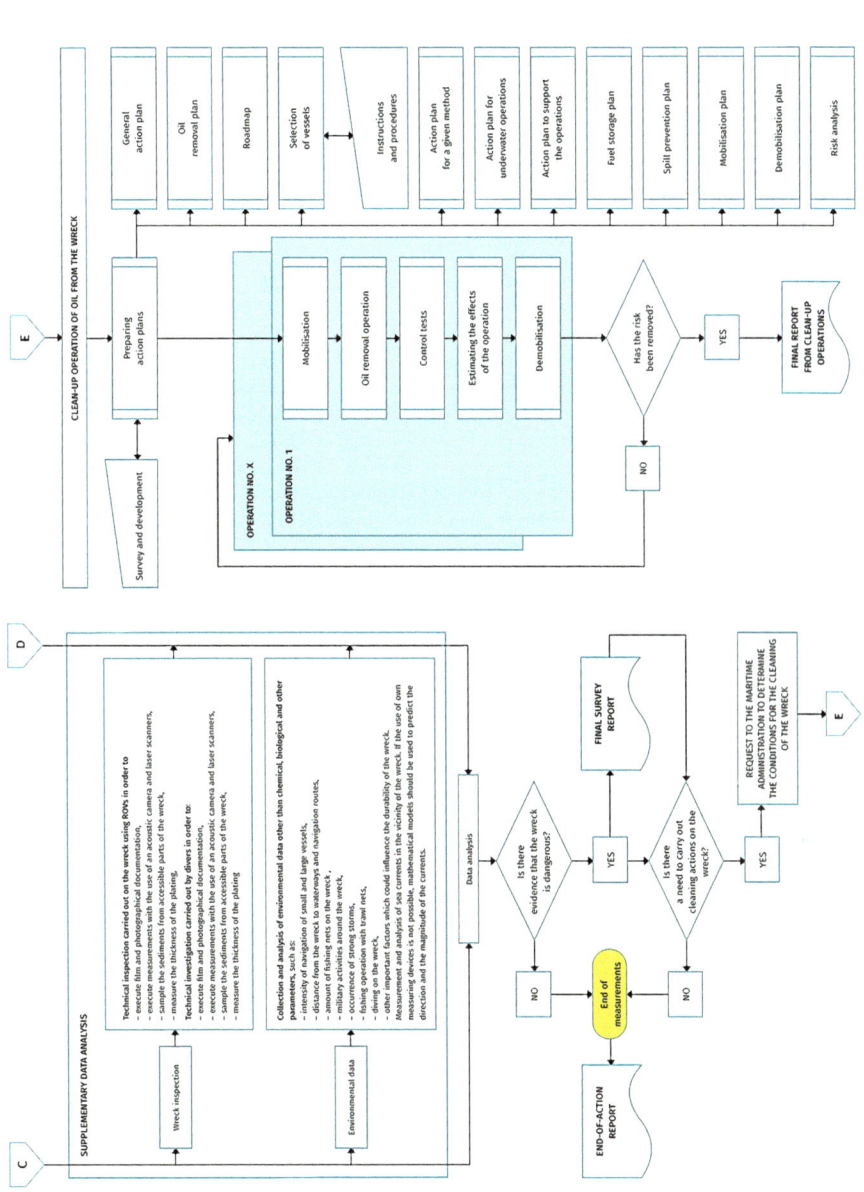

Fig. 7.5 Steps to be taken from the discovery of the wreck, through its identification and risk assessment, to the removal of fuel from the wreck—Part II (Hac & Sarna, 2021)

References

Baltic Sea. https://en.wikipedia.org/wiki/Baltic_Sea. Accessed 10 June 2023.

Bundesamt *für* Seeschifffahrt *und* Hydrographie Wrack Base (necessary access). https://www. bsh.de/EN/TOPICS/Surveying_and_cartography/Wreck_search/Definition_of_wreck_search/ definition_of_wreck_search_node.html; jsessionid=64E50AFD5D388CD7909B7B22310BB6 3D.live21323. Accessed 30 June 2023.

Danish Navy. (2013). The emptying of the Fu Shan Hai. 24th July—12th September 2013. 27 Slides.

Finnish Wreck Base. http://www.hylyt.net. Accessed 30 June 2023.

Hac, B. (2014). Hazard of seabed contamination by oil products from motor ship wrecks based on the example of the 'Stuttgart' ship wreck. Meeting of the HELCOM Submerged EWG, Szczecin 29–30.10.2014.

Hac, B. (2018). Preliminary action plan for the retrieval activities on the Franken shipwreck. Report from the research expedition carried out on the Franken shipwreck on 23rd-28th April 2018 In the framework of the project reduction of the negative impact of oil spills from the Franken shipwreck. The Mare Foundation and The Maritime Institute in Gdańsk.

Hac, B., & Sarna, O. (2021). General methodology of oil removal operations on Baltic shipwrecks. Proposition of a wreck management program for Poland. Report of The Mare Foundation. ISBN 978–83–959773-1-2. 110 p.

Jolma, K. (2009). Kokonaisselvitys valtion ja kuntien öljyntorjuntavalmiuden kehittämisestä 2009–2018. (Total assessment to develop oil combating preparedness of the government and municipalities—in Finnish). Finnish Environment Institute (SYKE). 49 s. 27.1.2009.

Michel, J., Etkin, D., Gilbert, T., Waldron, J., Blocksidge, C., & Urban, R. (2005). *Potentially polluting wrecks in marine waters: An issue paper presented at the 2005 international oil spill conference* (76 pp). American Petroleum Institute.

Mortensen, O., & Rasmussen, F. K. (2013). Bjaerning af olie fra FU SHAN HAI. Endelig rapport (in Danish) 19 pp.

Pärt, S., Kõuts Rytkönen, J. Rintamaa, J., Landquist, H. (2015). Report on wreck classification, potential for oil pollution. Report D 1.1. of the EU-BONUS-project SWERA (SWERA, Sunken Wreck Environmental Risk Assessment).

Rogowska, J., Wolska, J. (2019). The threat to the marine environment resulting from the presence of world war II shipwrecks, on the example of the s/s Stuttgart. Department of Environmental.

Rytkönen, J. (1999) .Uponneiden alusten öljypäästövaara (the threat of oil pollution from wrecks – in Finnish). Research report: VAL34-992239. VTT manufacturing technology, April 1999, Espoo, Finland.

Rytkönen, J. (2021). Wreck oil removal—Authorities and other parties competence development—Proposal on the development of cooperation between authorities (in Finnish). Finnish environment institute (SYKE). Report of the water protection Programme. 20.04.2021. https:// ym.fi/en/water-protection-programme.

SHIPWHER. (2010). *Wreck register*. http://register.muinas.ee/?menuID=en_wreckregistry &page=1&_nocache=1408705748; *https://daivings.lv/en/latvia-sunk-ship-wreck-map/*. Accessed 30 June 2023.

Swedish Maritime Administration. (2014). Environmental risks posed by sunken wrecks. Case No.: 1399-14-01942-6. This report is available on the website of the Maritime Administration: https://www.google.com/url?sa=t&source=web&rct=j&opi=89978449&u rl=https://research.chalmers.se/publication/206812/file/206812_AdditionalFile_997eb949. pdf&ved=2ahUKEwjSnO2-oLWFAxULJRAIHZHHAHEQFnoECA4QAQ&usg=AOvVaw3I bihryZWfVmsWR1s2QtE3

Svensson, E. (2010). Potential shipwreck pollution in the Baltic Sea. Overview of work in the Baltic Sea states. Report by Lighthouse & Swedish Maritime Administration 09-02375. 15.12.2010. 35 pp.

The Baltic Times. Wreck sunk over a century ago, found leaking fuel in Baltic Sea https://www.baltictimes.com/wreck_sunk_over_a_century_ago__found_leaking_fuel_in_baltic_sea/. Accessed 07 July 2023.

VRAK Museum of wrecks. https://www.vrak.se/en/wrecks-and-remains/skytteren/. Accessed 17 Sept 2023.

Open Access This chapter is licensed under the terms of the Creative Commons Attribution 4.0 International License (http://creativecommons.org/licenses/by/4.0/), which permits use, sharing, adaptation, distribution and reproduction in any medium or format, as long as you give appropriate credit to the original author(s) and the source, provide a link to the Creative Commons license and indicate if changes were made.

The images or other third party material in this chapter are included in the chapter's Creative Commons license, unless indicated otherwise in a credit line to the material. If material is not included in the chapter's Creative Commons license and your intended use is not permitted by statutory regulation or exceeds the permitted use, you will need to obtain permission directly from the copyright holder.

Chapter 8
Potentially Polluting Wrecks in the Blue Pacific

Matthew Carter, Ashley Meredith, Augustine C. Kohler, Bradford Mori, and Ranger Walter

8.1 Introduction to Potentially Polluting Wrecks in the Pacific

Marine pollution is a global and transboundary issue that negatively affects the environment, people, and coastal economies around the world. It is widely recognised as one of the four major threats to the world's oceans, along with climate change, habitat destruction and over-exploitation of living marine resources (SPREP, 2020). Since the large-scale introduction of oil-fueled shipping in the 1930s, marine pollution incidents have occurred in practically all coastal waters across the globe. In the Pacific Region the greatest concentration of these incidents stemmed from the loss of over 3800 ships during the World War II (WWII) (SPREP, 2003a, b) (Fig. 8.1). At the time of their loss, these ships contained as many as 1.5 billion gallons of petrochemicals, and hundreds of thousands of tons of explosive ordnance (Michel et al., 2005), an unknown amount of which remains in these wrecks today.

In the 'Blue Pacific', the health of the ocean is fundamental to the sustainability of all aspects of island life. The importance of coastal and marine environments to every aspect of the lives of Pacific islanders cannot be overstated, and the impacts

M. Carter (✉)
The Major Projects Foundation, Newcastle, Australia
e-mail: Matt.carter@majorprojects.org.au; m.j.carter@latrobe.edu.au

A. Meredith
National Cultural Anthropologist, FSM Office of National Archives, Culture and Historic Preservation (NACH), Pohnpei, Federated States of Micronesia

A. C. Kohler
FSM Office of National Archives, Culture and Historic Preservation (NACH), Pohnpei, Federated States of Micronesia

B. Mori · R. Walter
Chuuk Environmental Protection Agency, Chuuk, Federated States of Micronesia

© The Author(s) 2024
M. L. Brennan (ed.), *Threats to Our Ocean Heritage: Potentially Polluting Wrecks*, SpringerBriefs in Underwater Archaeology,
https://doi.org/10.1007/978-3-031-57960-8_8

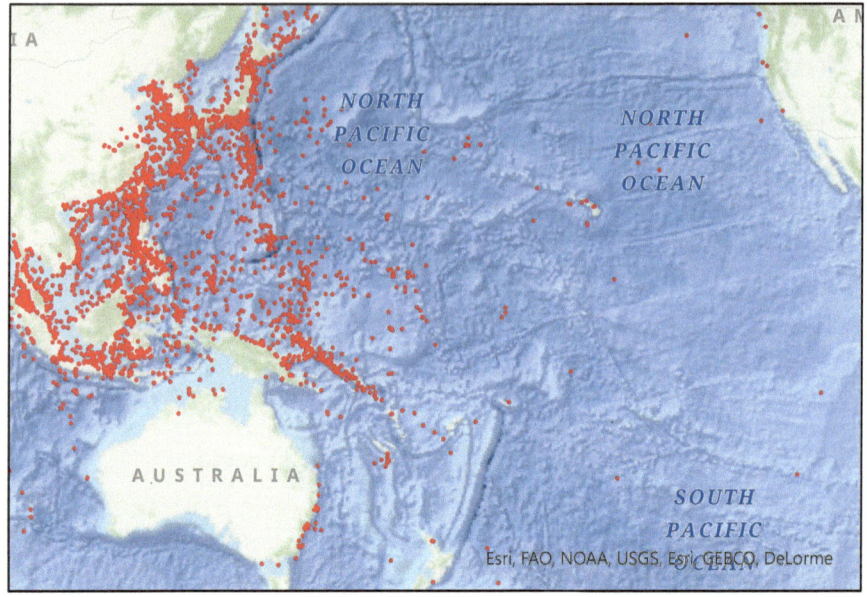

Fig. 8.1 Map illustrating the locations of WWII shipwrecks in the Asia Pacific region. (Attribution: © Paul Heersink, 2022)

of marine spills including those from shipwrecks constitute a major concern for Pacific Island peoples. Significantly, despite initiatives to assist Pacific Island countries and states with marine spill prevention and response, many of these nations are underprepared for the impending spills from the potentially polluting wrecks (PPW) within their waters.

The Pacific War was the largest maritime endeavor in human history. In 2003 the Secretariat of the Pacific Regional Environment Programme (SPREP) commissioned the creation of the 'SPREP Pacific WWII Shipwreck Database' which contained entries for 3855 vessels totaling over 13 million tons, ranging from aircraft carriers to battleships and including over 330 tankers and oilers. This database included the country of origin and therefore the ownership of the sunken vessels with 14 nations represented across both military and civilian merchant vessels. Significantly, of the 3855 vessels, 3326 are Japanese (86%), 415 are United States of America (10%), with the other 12 nations making up just 4% of the total (SPREP, 2003a).

While the exact volume of oil still held by the 3855 WWII vessels wrecked across the Pacific is unknown, researchers have estimated that this could range from between 190 million and 1.5 billion gallons (Michel et al., 2005). It is expected that the majority of oil within these WWII wrecks will be heavy fuel oil along with quantities of diesel, lubrication oils, and smaller amounts of aviation fuels and gasoline (Talouli et al., 2009). In the majority of the larger vessels sunk during the Pacific conflict, the fuel oil consisted of a mixture of both non-persistent and persistent oils. This combination appears to consist of bunker oil (No. 6 fuel oil) and marine diesel (No. 2 fuel oil) (Talouli et al., 2009). A significant factor in the evaporation of leaking fuel in the tropical Pacific is the high water and ambient temperatures of the region. Average Pacific air and water temperatures are high between 25–32 C degrees with often less than 10 C degree variance between night and day. This combined with predominantly windy conditions would accelerate the evaporation rate of the spilled fuel compared to cooler less windy locations.

As described by Talouli (et al., 2009), under the spill conditions in most central and southern Pacific locations, if released, the heavier bunker oil from WWII wrecks would behave in a manner similar to conventional #6 fuel oils. This heavy oil has a slightly lower density than full-strength seawater at tropical temperatures and is likely to float and remain liquid during the early stages of a spill. Some components will dissolve into the water, and the light fractions will be lost by evaporation, but what remains floating will initially form contiguous slicks. Eventually the slicks will break up into widely scattered fields of pancakes and tar-balls, which can persist over large distances and concentrate in convergence zones. The heavier persistent components can form emulsions in rough seas or manifest as tar-balls on coastlines, or even travel great distances across the open sea. Because of the higher viscosities of these oils, the tar-balls may be more persistent than expected for conventional crude oils, potentially reaching intertidal zones if close to the shore and settle onto sediments.

In contrast to the heavy fuel, diesel oil leaking from WWII wrecks would be expected to undergo rapid weathering, with the majority of any diesel spilled in tropical Pacific waters being either dispersed into the water column or evaporated within a span of 12–24 h. This does not negate the potential ecological impacts on aquatic life, coral reefs, or potential effects on wildlife, however. It indicates that the diesel oil would be relatively quickly removed from the water surface following its release. Once dispersed or dissolved within the water column, diesel oil could still exert notable effects on intertidal life and fisheries, while also being harder to detect or clean up.

The impacts of oil and diesel spills from PPW in the Pacific are wide ranging and threaten marine ecosystems, cultures and livelihoods across the region (SPREP, 2019). As noted by Talouli (et al., 2009) such impacts range from:

- The modification of natural habitats through both physical and chemical means,
- The smothering of fauna and flora.
- Both lethal and sub-lethal toxic impacts on fish, as well as other wildlife and plant life.
- Short-term and long-term shifts in biological communities
- Contamination of edible species, particularly fish and shellfish.
- The inability to utilise recreational areas like sandy beaches.
- Decline in demand for fisheries and tourism.

- Contamination of boats, fishing equipment, boat ramps, jetties, and related infrastructure.
- Temporary halt in operations for industries reliant on the marine environment.

The vast majority of PPW across the Pacific date from WWII and have been submerged for at least 79 years. Over this period, it is estimated that they have experienced a corrosion rate of approximately 0.1 mm per year, putting them at an increasing risk of structural failure (Melchers, 2003, 2013; Macleod, 2016). The ongoing corrosion of these PPW represents a critical challenge in the sustainable management of these sites, as the wrecks are steadily approaching inevitable collapse releasing any remaining pollutants they contain. Corrosion surveys undertaken on the PPW in Chuuk Lagoon in the Federated States of Micronesia revealed that although these wrecks are corroding at a rate 26–30% slower than wrecks in the open ocean at similar depths (MacLeod, 2003), they are projected to experience significant collapse within the next 3 years (2023–2026), leading to the release of oil into the surrounding marine environment (Macleod et al., 2011; Macleod, 2016). Furthermore, this corrosion and subsequent damage are likely to be exacerbated by more severe weather events associated with climate change (Macleod et al., 2017), with reports already indicating partial collapse in several wrecks (Aisek G 2023, pers. comm., 3 November).

8.2   History of PPW Management in the Pacific

The first formal recognition of the threat of PPW in the Pacific came from the Solomon Islands Government in 1999 when they requested the South Pacific Applied Geoscience Commission (SOPAC) to conduct a contamination risk assessment of WWII sunken ships and aircraft in Iron Bottom Sound, Solomon Islands (Maharaj, 1999). This study funded by the United Nations Development Programme found that the wrecks 'represent a real source of pollution to the natural environment' through the leakage of oils and fuel, leaching of trace element and heavy metals from paints, corroded aircraft and ships and munitions. The study put forward recommendations for future work to assist in the sustainable planning and management of these PPW. This recommendation was furthered by the Secretariat of the Pacific Regional Environment Programme (SPREP) who proposed a project to remove the oil from all WWII shipwrecks in the Solomon Islands 'where feasible and practicable' (SPREP, 1999). Unfortunately, neither the recommendations put forward by SOPAC nor the oil removal project proposed by SPREP were eventuated.

At the 12th assembly of the South Pacific Regional Environment Programme (SPREP) in September 2001, the USS *Mississinewa* incident (described below), prompted member countries to request a regional plan for addressing marine pollution from WWII shipwrecks. This responsibility fell to the Pacific Ocean Pollution

Prevention Programme (PACPOL) within SPREP. A preliminary version of the Regional Strategy was formulated and presented during the 13th SPREP Meeting in July 2002 (Nawadra, 2002). The subsequently approved Strategy encompassed five steps, covering the identification and characterisation of wrecks, conducting a Generic Risk Assessment Ranking, and agreeing on Intervention Assistance (Steps 1 to 3). It also included specific Site Based Risk Assessment and Remedial Action Implementation (Steps 4 and 5) (SPREP, 2003b). This strategy was subsequently advanced through the creation of a database identifying 3855 WWII shipwrecks in the Asia-Pacific region (SPREP, 2003a). Despite the evident advantages of approaching the issue of PPWs as a multilateral regional group, during the 14th SPREP Meeting in Apia, Samoa, in September 2003, it was decided that the threat of PPWs would instead be addressed through bilateral cooperation between the Coastal State and the Flag State, with SPREP offering assistance upon request. Subsequent to this decision information about these PPWs has been further refined (Michel et al., 2005; Monfils, 2005; Monfils et al., 2006). However, regional progress has been slow, with much of the work conducted through bilateral agreements in an ad hoc manner (Talouli et al., 2009).

8.3 U.S Government PPW Interventions in the Pacific

As noted above, 415 U.S ships were sunk in the Pacific accounting for 10% of the WWII Pacific losses. Despite this relatively large number of wrecks, the US Government has only acted to remove oil from three legacy wrecks across the region.

8.4 USS *Mississinewa*

Between July and August 2001, the wreck of the United States Navy oiler USS *Mississinewa* (1943–1944) leaked approximately 24,000 gal of oil into Ulithi Lagoon, on the island of Yap in the Federated States of Micronesia (FSM) (Gilbert, 2001; Gilbert et al., 2003). In response, a state of emergency was declared in Yap and a ban on all fishing activities in the affected area was enforced. Significantly, the spill led the FSM Government to request assistance from SPREP under the Pacific Islands Regional Marine Spill Contingency Plan (PACPLAN) adopted in October 2000 (Gilbert, 2001). This request was the first of its kind and saw SPREP's Marine Pollution Advisor dispatched to Yap to compile an independent study of the wreck and the environmental impacts of the recent oil spill. This assessment concluded that the wreck's extensive oil cargo posed an unacceptable and ever-present risk to the marine environment of Ulithi Lagoon and should be offloaded as a permanent solution (Gilbert, 2001). Following this report the Governor of Yap contacted the

U.S. Embassy in FSM and requested that the remaining oil from USS *Mississinewa* be removed. In response the Embassy initiated discussions with other federal agencies, including the U.S. Coast Guard (USCG) District 14 (Honolulu) and USCG headquarters (Washington), the National Oceanic and Atmospheric Administration (NOAA), the Environmental Protection Agency (EPA), and the Department of the Interior (DOI). Following these discussions, the U.S. Navy conducted a substantial operation to evaluate the wreck and execute plans for oil extraction. This undertaking required the deployment of various vessels, barges, and personnel, resulting in salvage costs of around $6 million USD (Naval Sea Systems Command, 2003). Between January and March 2003, all readily accessible oil was successfully removed from the vessel. It is estimated that approximately 1.8 million gallons of oil and diesel fuel were successfully recovered. Interestingly, this fuel was subsequently shipped to Singapore where it was sold.

8.5 Ex-USS *Chehalis*

In October 1949 ex-USS *Chehalis* (a WWII-era gasoline tanker) sank in Pago Pago Harbor, American Samoa. In 2007 after increasingly frequent reports of oil and fuel leaks coming from the vessel a commercial survey team conducted an inspection. Their findings indicated that approximately 40,000 gallons of motor gasoline and 70,000 gallons of aviation gasoline were still present onboard. Based on this survey, American Samoa and the U.S. Environmental Protection Agency (EPA) jointly petitioned the U.S. Coast Guard (USCG) to take necessary measures for the removal of the remaining fuel from ex-USS *Chehalis*. In March 2010, a multiagency effort involving the US Coast Guard, US Navy Supervisor of Salvage (SUPSALV), the National Oceanic and Atmospheric Administration (NOAA), National Pollution Fund Center, and the EPA commenced operations to safely remove fuel from the vessel. This endeavor led to the extraction of 54,505 gallons of fuel, which was subsequently transported to Systech Environmental Corporation in Fredonia, Kansas where it underwent recycling in their cement-making kiln.

8.6 Ex-USS *Prinz Eugen*

In 2017, the U.S. Indo-Pacific Command (USINDOPACOM) enlisted the services of SUPSALV to conduct an extensive diver survey of the sunken WWII-era, former German cruiser, ex-USS *Prinz Eugen*, situated in Kwajalein Atoll within the Republic of the Marshall Islands (RMI). The primary objective of this initial survey was to assess the current condition of the hull and gather essential data regarding the presence and volume of any remaining oil in the vessel's tanks, as there was growing evidence of oil seepage from the wreck. Subsequent to the survey, it was determined that a removal operation was required to mitigate the potential for a significant oil release from the deteriorating wreck as such an event could have adverse effects on the marine ecosystem, the neighboring human population, and U.S. Army property.

Given that the Republic of the Marshall Islands government possessed ownership rights to the wreck, the U.S. government needed to obtain both approval from the Republic of the Marshall Islands and authorisation from Congress to allocate U.S. funds for the disposal of a foreign-owned vessel. A Diplomatic (DIP) Note was executed, establishing an accord between the Government of the Republic of the Marshall Islands (GRMI) and the U.S. Government (USG). This agreement waived the environmental prerequisites for extensive assessments, consultations, and permitting as outlined in the 1986 Compact agreement. It also indemnified the USG from any liabilities arising from the ex-USS *Prinz Eugen* and granted the USG permission to extract and properly dispose of the oil from the wreck, despite the Republic of the Marshall Islands' ownership. In October of 2018, a collaborative team led by the U.S. Navy successfully completed the extraction of 229,000 gallons of oil from 173 fuel tanks on the wreck.

8.7 Japanese Government Interventions in the Pacific

Despite the vast majority of PPW dating from WWII in the Pacific being of Japanese origin (86%), the Japanese Government has not formally undertaken any proactive steps to address the risk these PPWs pose. Instead, the Japanese Government, through their Ministry of Foreign Affairs, has funded a small number of projects undertaken by the Japanese Mine Action Service (JMAS), a Japanese NGO on PPW in Palau and Chuuk.

The work of JMAS in Palau began in December 2012 and involved investigating the underwater explosive remnants of WWII by former members of the Japan Maritime Self Defence force employed by JMAS. In relation to PPW this work mainly revolved around the recovery and safe destruction of depth charges from a shipwreck locally known as the 'Helmet Wreck'. In 2015 the depth charges were found to be leaking their contents, an explosive chemical known as picric acid, which was subsequently polluting the surrounding waters (UNESCO, 2017). In 2019 JMAS partnered with the Norwegian People's Aid to begin removing and destroying the 165 depth charges on the wreck, with this work continuing into 2022. While no report has been found, it is believed that the JMAS team also undertook some surveys of PPW in Palau and are reported to have identified at least one wreck where oil was leaking (Takagi M. 2023, pers. comm., 2 June).

In 2017 JMAS began the 'Oil Leakage Countermeasures from WWII Wrecks of FSM Truk Lagoon Marine Area' project. This work was funded through a financial grant of USD $857,899 from the Government of Japan and was scheduled to run for 3 years from June 2017 to June 2020. This project was designed around five specific activities namely, (1) identifying the location of the target wrecks, (2) generating 3D images of the wrecks, (3) clarifying the current condition and cargo of the wrecks, (4) understanding the oil leak conditions of the wrecks and (5) implementing measures to prevent oil leaking and to remove oil (Inoue K. 2019, pers. comm., 20 July). This work identified that approximately 43,000 gallons of oil remained trapped inside the hulls of the 11 wrecks that were surveyed by the JMAS team. In September 2021 the project was extended with an additional USD $741,206 provided by the Government of Japan. A key priority of the JMAS work was removing pockets of oil trapped inside the wrecks and between May 2017 and June 2023 JMAS divers were able to remove approximately 10,000 gallons of oil from five of the 11 wrecks (Takagi M 2023, pers. comm., 10 June). Unfortunately, no provision was made in the project for funding the export and disposal of the collected oil from Chuuk and it is currently being stored in a Government warehouse while the quoted USD $107,000 required to export it is raised by the Chuuk or FSM Government (Mori B. 2023, pers. comm., 31 October). In June 2023 JMAS withdrew their team from Chuuk and presently it is unclear if they will return. The JMAS mission has started to reveal the scale of the threat these PPW pose and has highlighted the urgent need for increased collaborative action in Chuuk to address the rest of the wrecks, and the oil sources beyond the reach of JMAS's capabilities.

8.8   Current Situation of PPW in the Pacific

While the remediation of USS *Mississinewa,* ex-USS *Chehalis* and ex-USS *Prinz Eugen* were successful, and the JMAS projects in Palau and Chuuk represent practical contributions, the threat PPWs pose continues to be identified as a key issue by Pacific Leaders at international assemblies. In 2019 ministers from 21 Pacific countries reaffirmed their recognition of the 'significant environmental threat' that the wrecks of WWII ships still pose to the Pacific (SPREP, 2019) and acknowledged the need to address PPW with an agreed regional approach that addresses both environmental conservation and community resilience. This message was again reiterated by the Pacific Islands Forum at the 2020 United Nations Summit on Biodiversity where the Prime Minister of Tuvalu called for 'the reduction and elimination of any threat posed to our people and ecosystems by pollution including nuclear waste, radioactive and other contaminants, shipwrecks and World War II relics'. In September 2023 the Director General of SPREP speaking at the international oil spill conference for the Asia-Pacific region once again reiterated the need for a proactive multilateral approach to mitigating the threat that PPWs pose to the region.

8.9 Case Study: Potentially Polluting Wrecks in Chuuk Lagoon

Chuuk Lagoon, formerly known as Truk Lagoon, is an atoll located in the central Pacific. It is situated approximately 1800 km (about 970 nautical miles) northeast of New Guinea and is a part of Chuuk State within the Federated States of Micronesia (FSM). The lagoon is encircled by a protective reef that spans 225 km (about 140 miles) and encompasses a natural harbor measuring 79 by 50 km (about 43 nautical miles by 27 nautical miles), with a total area of 2130 km² (about 820 square miles). Chuuk Lagoon is inhabited by around 53,000 Chuukese people, making it the most densely populated region in the Federated States of Micronesia.

During World War II, it served as a significant Japanese base for their combined fleet and played a crucial role as a strategic advance base for Japan's expansion into the southern Pacific. From February 1944 to August 1945, the U.S. Navy conducted extensive bombing campaigns, resulting in the destruction of airstrips, military facilities, and the sinking of over 50 ships. This effectively neutralised the base and played a pivotal role in the outcome of the war. Today, Chuuk Lagoon contains one of the largest concentrations of potentially polluting shipwrecks identified in the Pacific Ocean; 19 of the wrecks have been identified as posing significant environmental risks due to the volumes of toxic fuel oil and unexploded ordnance they hold (Carter et al., 2021) (Fig. 8.2).

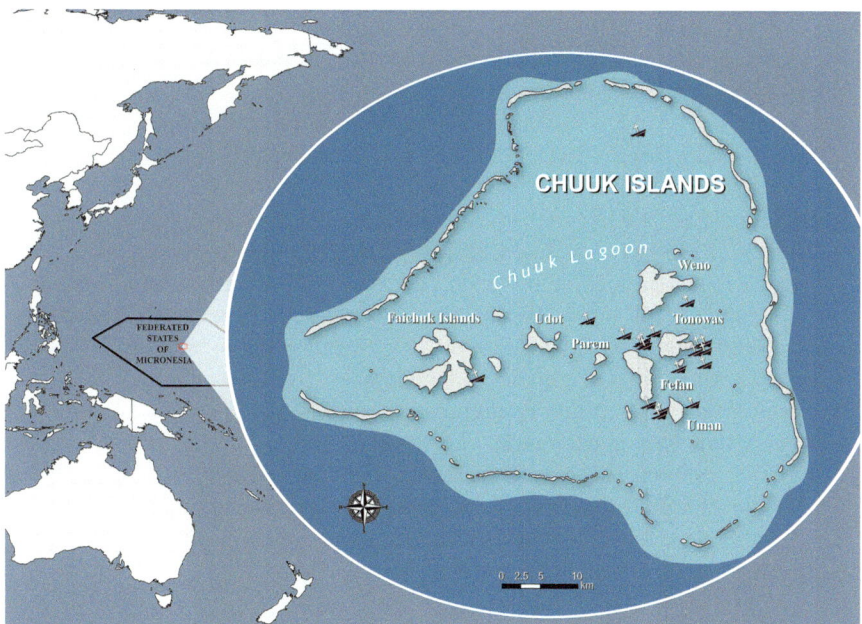

Fig. 8.2 Location map of Chuuk Lagoon showing the location of the PPW identified as posing significant environmental risks

In 1976, Earle and Giddings (1976: 603) first brought to light the issue of fuel leakage from a sunken vessel within the Lagoon, specifically emphasising the case of the *Amagisan Maru*. Over the next 30 years in addition to diesel and oil fuel, reports indicated that aviation gasoline was also detected seeping from drums on the armed merchant transport *Kiyosumi Maru* (Hodson, Colin 2001; pers. comm. Bill Jeffery). Subsequently, in July 2006, aerial surveys revealed that the passenger-cargo ship *San Francisco Maru* and the tanker *Hoyo Maru* were experiencing leakage (Osiena, R. Chuuk Director Department of Marine Resources, 2006 pers. comm. to Bill Jeffery). Over the years, several other sites have experienced fuel and oil leaks, with surface slicks being utilised by Chuukese dive guides to aid in locating these sites.

In 2009, reports emerged of oil seeping from the wreck of *Hoyo Maru*, prompting investigations by SPREP and the Chuuk EPA into nearby beaches for evidence of oil contamination. A subsequent report stemming from this investigation concluded that the PPW in Chuuk 'pose a serious and immediate threat to the people, marine and coastal environments, tourism, and fisheries of the region' (Talouli et al., 2009). Building on this assessment, in 2011, the President of Micronesia made an urgent appeal to the United Nations General Assembly for immediate assistance in addressing the WWII shipwrecks in Chuuk Lagoon, characterising them as 'ticking environmental time bombs'. Despite this plea no support was forthcoming and in 2014 the issue was once again raised by Mr. Andrew Yatilman from the FSM Office of Environment and Emergency Management (OEEM), emphasising the critical need for assistance from partners and donors. He stressed the urgency of conducting a comprehensive assessment to ascertain the status, risks, and associated costs of retrieving the oil from those wrecks that posed, or had the potential to cause, the most significant harm.

On April 19, 2018, FSM made history by becoming the first country in the Pacific to ratify the UNESCO Convention on the Protection of the Underwater Cultural Heritage 2001 (UCH Convention). This convention is recognised as international best practice in managing underwater cultural heritage. Through this ratification, FSM has committed itself to a legally binding framework aimed at enhancing the identification, research, and safeguarding of their underwater heritage, including PPW, all while ensuring its long-term preservation and sustainability. While the UCH Convention provides overarching protection for all UCH items over 100 years old, it does not impede states from safeguarding historically, archaeologically, or culturally significant UCH items that are younger than 100 years old. This is exemplified by the case of the WWII wrecks in Chuuk, where both national and state legislations have been enacted to safeguard these wrecks, despite their age being less than a century. The stipulation is that any activities directed towards these UCH sites must adhere to the rules and annex outlined in the 2001 UNESCO Convention.

8.10   The FSM WWII Shipwreck Pollution Mitigation Project

In July 2021 the ninth Pacific Islands Leaders Meeting (PALM), co-chaired by Prime Minister Suga Yoshihide of Japan and Prime Minister Kausea Natano of Tuvalu, emphasised the importance of protecting the ocean from harmful plastics and threat posed by the presence of nuclear waste, radioactive and other contaminants, shipwrecks and World War II relics. As a result of this call the Government of Australia (GOA) offered assistance to FSM to support the efforts of Japan and FSM to address the risks from leaking PPW in Chuuk Lagoon. The assistance from the GOA took the form of a commitment of USD $1.38 million to partner with SPREP and the Major Projects Foundation (MPF) to complement and enhance the work underway by JMAS in Chuuk Lagoon. This project began in March 2022 and will run for 3 years with the intention of achieving the following outcomes.

1. Reduce the likelihood and impacts of oil spills from PPW in Chuuk Lagoon by accelerating and enhancing the efforts of JMAS in removing trapped oil within wrecked vessels.
2. Strengthen the FSM's capacity to manage the threat of PPW in Chuuk Lagoon by providing technical information.
3. Enhance the contingency response capabilities of the FSM and Chuuk authorities to address oil pollution from PPW.

Due to the closure of the FSM borders during the COVID 19 pandemic, the project team was unable to visit Chuuk until November 2022. At this time work began between JMAS and MPF to improve the procedures for removing oil from the PPW in Chuuk and also to undertake the technical survey of the target wrecks.

While plans were being developed to further enhance the JMAS work their withdrawal from Chuuk in June 2023 has required a redesign of the project. Subsequent fieldwork by MPF in Chuuk in November 2023 focused on assessing the practicality of resuming oil removal operations and continuing with the technical survey of the wrecks. Significantly, the FSM WWII Shipwreck Pollution Mitigation Project is currently the only work being undertaken proactively on legacy PPW in the Pacific. The intention is for this project to provide a best practice template that can be utilised by other Pacific nations in their management of PPW within their waters.

8.11   Lessons and Conclusions

The threat that PPW pose to the Pacific has been recognised since at least the late 1990s with subsequent research focusing on identifying the scale of the threat (3855 wrecks), and the potential timing of their collapse (imminent). Organisations such as SPREP have become acutely aware of the limited capacity for Pacific Island countries and Territories to respond to spills from PPW and note the impacts of these spills as being particularly devastating in a region so heavily intertwined with the ocean.

Despite repeated and urgent calls from the Pacific, both the U.S. (with 10% of the total PPW) and Japan (with 86%) remain committed to a reactive approach to this issue. When forced, the U.S. Government has invested substantial funds and expertise in removing oil from three PPWs in the region, but it has yet to adopt a systematic proactive approach to dealing with its remaining PPW. Similarly, the funding provided by the Japanese Government for work on their PPW in Palau and Chuuk falls far short of what is needed, especially in light of their large number of PPW across the Pacific Region.

Despite limited progress in addressing the PPW threat in the Pacific, several key issues have become apparent. Most legacy PPW fall under the principle of 'sovereign immunity,' belonging to the government in control of the vessel at the time of its sinking (Monfils et al., 2006). However, what remains unclear is who is responsible or authorised to remove pollutants from these wrecks, a question that has yet to be satisfactorily answered in the Pacific. Another challenge is what to do with the pollutants once they are removed. Pacific Island countries, including the Federated States of Micronesia, Solomon Islands, Republic of the Marshall Islands, and Papua New Guinea, already struggle with waste oil, and this situation would be greatly exacerbated by the addition of pollutants from PPW. Additionally, there is a significant cost associated with exporting waste oil from these countries for proper disposal, along with the need to ensure compliance with the Basel and Waigani Conventions regarding the export and import of hazardous waste.

The threat posed by PPW to the marine ecosystems, cultures, and livelihoods of the Pacific is evident and growing more pressing each day. The sheer scale of the problem and the complex challenges it presents necessitate a multilateral solution. Countries responsible for PPW must collaborate with impacted nations to secure the funding and expertise required to mitigate the impact of the region's PPW and safeguard the Blue Pacific for future generations.

References

Carter, M., Goodsir, F., Cundall, P., Devlin, M., Fuller, S., Jeffery, B., & Talouli, A. (2021). Ticking ecological time bombs: Risk characterisation and management of oil polluting World War II shipwrecks in the Pacific Ocean. *Marine Pollution Bulletin, 164*, 112087.

Earle, S. A., & Giddings, A. (1976). Life springs from death in Truk Lagoon. *National Geographic, 149*(5), 578–612.

Gilbert, T. (2001). *Report of the strategic environmental assessment.* USS Mississinewa Oil Spill.

Gilbert, T., Nawadra, S., Tafileichig, A., & Yinug, L. (2003). Response to an Oil Spill from a Sunken WWII Oil Tanker in Yap State, Micronesia. *International Oil Spill Conference Proceedings, 2003*(1), 175–182.

Macleod, I. D. (2003). Metal corrosion in Chuuk Lagoon: A survey of iron shipwrecks and aluminium aircraft (Western Australian Museum).

Macleod, I. D. G. (2016). In-situ corrosion measurements of WWII shipwrecks in Chuuk Lagoon, quantification of decay mechanisms and rates of deterioration. *Frontiers in Marine Science, 3*(38), 1–10.

Macleod, I. D. G., Richards, Z. T., & Beger, M. (2011). The effects of human and biological interactions on the corrosion of WWII iron shipwrecks in Chuuk Lagoon. In *18th International Corrosion Congress 2011*, 1, pp. 298–309.

Macleod, I. D. G., Selman, A., & Selman, C. (2017). Assessing the impact of typhoons on historic Iron shipwrecks in Chuuk lagoon through changes in the corrosion microenvironment. *Conservation and Management of Archaeological Sites, 19*(4), 269–287.

Maharaj, R. J. (1999). *Contamination risk assessment from WWII Armoury in Iron Bottom Sound, Solomon Islands*. Apia, Samoa, SOPAC.

Melchers, R. (2003). Modeling of marine immersion corrosion for mild and low-alloy steels–part 1: Phenomenological model. *Corrosion, 59*(4), 319.

Melchers, R. E. (2013). Long-term corrosion of cast irons and steel in marine and atmospheric environments. *Corrosion Science, 68*, 186–194.

Michel, J., Gilbert, T., Waldron, J., Blocksidge, C., Etkin, D.S., Urban, R. (2005). Potentially polluting wrecks in marine waters. In *2005 International Oil Spill Conference, IOSC 2005*.

Monfils, R. (2005). The global risk of marine pollution from WWII shipwrecks: Examples from the seven seas. *International Oil Spill Conference Proceedings: May 2005* 2005 (1), 1049–1054.

Monfils, R., Gilbert, T., & Nawadra, S. (2006). Sunken WWII shipwrecks of the Pacific and East Asia: The need for regional collaboration to address the potential marine pollution threat. *Ocean & Coastal Management, 49*(9), 779–788.

Naval Sea Systems Command. (2003). U.S. navy salvage report, USS *Mississinewa* oil removal operations. Naval Sea Systems Command, S0300-B6-RPT-010, 0910-LP-102-8809.

Nawadra, S. (2002). Regional Strategy to Address Marine Pollution from World War II Shipwrecks, Thirteenth SPREP Meeting of Officials (Item 7.2.2.1) Majuro, Marshall Islands 21–25 July, 2002. Apia, Samoa, SPREP.

SPREP. (1999). Pacific ocean pollution prevention programme (PACPOL): Strategy and workplan. Apia, Samoa, Secretariat of the Pacific Regional Environment Programme.

SPREP. (2003a). World War II ship wrecks of the Pacific – Strategy status of database. Apia, Samoa, SPREP.

SPREP. (2003b). Regional strategy to address marine pollution from World War II shipwrecks. In *Majuro, Marshall Islands*. Secretariat of the Pacific regional environment programme, Apia, Samoa.

SPREP. (2019). SPREP environment ministers and high-level representatives Talanoa communique 2019. Apia, Samoa, SPREP.

SPREP. (2020). PACPLAN: Pacific Islands regional marine spill contingency plan 2019. Apia, Samoa, SPREP.

Talouli, A., Gilbert, A., & Monfils, R. (2009). Strategic environmental assessment and potential future shoreline impacts of the oil spill from WWII shipwreck Hoyo Maru Chuuk Lagoon – Federated States of Micronesia. Apia, Samoa, SPREP.

UNESCO. (2017). *Safeguarding underwater cultural heritage in the Pacific: Report on good practice in the protection and management of World War II-related underwater cultural heritage*. UNESCO.

Open Access This chapter is licensed under the terms of the Creative Commons Attribution 4.0 International License (http://creativecommons.org/licenses/by/4.0/), which permits use, sharing, adaptation, distribution and reproduction in any medium or format, as long as you give appropriate credit to the original author(s) and the source, provide a link to the Creative Commons license and indicate if changes were made.

The images or other third party material in this chapter are included in the chapter's Creative Commons license, unless indicated otherwise in a credit line to the material. If material is not included in the chapter's Creative Commons license and your intended use is not permitted by statutory regulation or exceeds the permitted use, you will need to obtain permission directly from the copyright holder.

Chapter 9
Satellite Detection and the Discovery of *Bloody Marsh*

Michael L. Brennan, Geoffrey Thiemann, and William Jeffery

9.1 Introduction

Assessment and mitigation of potentially polluting wrecks in deeper water is a more challenging endeavor due to limitations of divers and operating pumping equipment at depth, but additionally because locating these wrecks is harder. While wrecks in deep water are more out of sight and out of mind, even when leaking, than those near shore, they are safer from damage by dredges, trawls or other anthropogenic activities. The wreck of the tanker SS *Bloody Marsh*, reportedly sunk off South Carolina on July 2, 1943 in 560 m of water, was considered a lower risk in NOAA's PPW study due to its location in deep water, which the NOAA Screening Level Risk Assessment Packages typically considered as less risk: 'deepwater shipwrecks tend to settle upright on the bottom, and is supported by the conclusions made by the U.S. Coast Guard in 1967 that oil will likely escape from a wreck's vents and piping long before its hull plates corrode' (NOAA, 2013b: 6). This, however, is a conclusion that needs to be reconsidered. *Bloody Marsh* was carrying a cargo of 106,496 barrels of bunker C heavy fuel oil, which is one of the largest cargoes among the vessels on the PPW list. In reviewing the 87 wrecks on the list, we selected *Bloody Marsh* as a high priority because of its large cargo and the fact that it was struck with two torpedoes and reportedly broke in half while sinking. Only ships with intact hulls would settle upright on the seabed, and therefore presumed *Bloody Marsh* did not.

M. L. Brennan (✉)
SEARCH Inc., Jacksonville, FL, USA
e-mail: mike@brennanexploration.com

G. Thiemann · W. Jeffery
CGG Services, Crawley, UK

© The Author(s) 2024
M. L. Brennan (ed.), *Threats to Our Ocean Heritage: Potentially Polluting Wrecks*, SpringerBriefs in Underwater Archaeology,
https://doi.org/10.1007/978-3-031-57960-8_9

This observation comes from a Coast Guard report from 1967 that indicates that wrecks that settle upright on the seabed will lose their oil through vents, and that ships that sink in deep water tend to settle upright (NOAA, 2013b: 6). While this is often the case due to the streamlined design of a vessel's hull, and hence a postulated hydrodynamic glide to the seabed, and some ships that have been sunk by torpedoes and lie in deep water have been found to settle upright, for example that of SS *Coast Trader* off Vancouver (Delgado et al., 2018), the assumption here is that vessels remained intact. *Coast Trader* was struck by a single torpedo and remained intact. *Bloody Marsh* was struck by two torpedoes and witness accounts indicated it broke in half on the surface. Therefore, when reviewing the list of PPW wrecks, we suspected that *Bloody Marsh* would be in sections on the seabed like *Coimbra* and *Munger T. Ball* (Brennan et al., 2023), and likely not upright. Another tanker wreck that illustrates this sinking pattern is that of RFA *Darkdale*, a British ship sunk of St. Helena island in the South Atlantic in 1941, which was broken into two halves, the bow inverted, and the stern section lying on its port side (Liddell & Skelhorn, 2012; Hill et al., Chap. 6 and Lawrence et al., Chap. 11, this volume).

We were able to locate *Bloody Marsh* at a depth of 465 m off South Carolina due to a fortunate survey of opportunity and that it was at a depth where hull-mounted multibeam sonar could detect it. This linked with reports of oil slicks on the surface nearby directed us to where to survey, which will be discussed in more detail herein. This wreck was also important to find due to its large amount of cargo (>106,000 barrels), and location near the Gulf Stream where a catastrophic leak could oil the shoreline from north Florida to Delaware (NOAA, 2013b: 31). However, due to the assumption about ships in deep water, the PPW study concluded: 'this is more likely a shipwreck to be aware of and not one considered for in-water assessment... it is unlikely that significant amounts of oil remain on the wreck' (2013b: 6).

Writing off wrecks in deep water due to the assumptions discussed above is premature, as the case of *Bloody Marsh* indicates. Numerous vessels were struck by multiple torpedoes and could lie in sections on the seabed, on their sides or inverted, positions in which oil would not escape through vents and piping. This example furthers the need for increased ocean exploration efforts to locate and assess potentially polluting wrecks and determine the exact state of the wreck site to progress past the desktop review initially put forth in the PPW study.

9.2   SS *Bloody Marsh*

SS *Bloody Marsh* was a T2-SE-A1 oil tanker built in Chester, Pennsylvania by the Sun Shipbuilding and Dry Dock Company in 1943 (Fig. 9.1). The T2 class tankers were a pre-fabricated commercial design that started production in 1940 by Sun Shipbuilding for the Standard Oil Company. These tankers were designed with standard dimensions and commonly available machinery to fast-track their construction. The T2-SE-A1 tanker was an 'oceangoing, single screw oil tanker of about 500 feet waterline length that had turboelectric propulsion, carried less than twelve

Fig. 9.1 *Bloody Marsh* (U.S. National Archives)

passengers, and had a deadweight of 16,000 tons (approximately 141,000 barrels)' (Spyrou, 2006: 3). Following the attack on Pearl Harbor in December 1941, the U.S. Maritime Commission ordered additional tankers to supply the war effort and replace those being lost from U-boat attacks. A total of 536 T2 tankers were built of three types, and 525 were nearly identical T2-SE-A1 vessels. This basic type vessel was based on the standard ship designed and built by the private commercial oil company in response to the establishment of an emergency tanker program, which was in response to the signing of the Lend-Lease Act in March 1941 to provide aid to Britain (Sawyer & Mitchell, 1974). The hulls of these tankers were all welded, which was a relatively new method introduced in the late 1930s. Some of these tankers developed hull fractures due to the electric welds and quality of the mild steel used for the hulls in cold temperatures that made the crystalline structure of the metal brittle (Spyrou, 2006). Notable instances of these tankers breaking in half include *Schenectady* in January 1943 and SS *Pendleton*, which sank in a storm off Cape Cod in February 1952 and inspired the novel and film, *The Finest Hours*. *Bloody Marsh* was sunk in warm summer waters off South Carolina on its maiden voyage, so the structure of its hull did not have the chance to be tested.

Bloody Marsh was loaded in Houston with 106,496 barrels of Navy fuel with specific gravity 13, and departed for New York under Master Albert Barnes operating for the Cities Service Oil Company. The tanker was not completely full; reports indicate tanks 2–8 were full as were wing tanks 3–8, and wing tank #2 was approximately ¾ full. Not sailing with a convoy, the tanker sailed from Houston in a zigzag pattern through the Gulf of Mexico and around Florida, as was common anti-submarine defense procedure. About 75 min before the attack, the tanker stopped this pattern and ran a straight course (Burch, 1943). The first torpedo from U-66 struck the port quarter at the engine room, causing the ship to immediately begin to settle by the stern, and the ship sunk until the gun platform was 'nearly awash' when the second torpedo struck amidships at the #3 tank, breaking the ship in two. The stern sank almost immediately while the bow floated free for 3–5 min before sinking (Burch, 1943: 2–3). The torpedo indicator went off about 30 s prior to when the first torpedo struck. The ship turned hard to port and was swinging in that direction as the torpedo hit. The bridge issued a distress signal 'SOS SSSS SOS BLOODY MARSH 31.33 N 78.55 W TORPEDOED' two or three times following the first strike, which was the only transmission the tanker sent during its entire voyage (Burch, 1943: 6). The course steadied after the explosion and gradually slowed, then proceeded at about two knots until the ship was abandoned.

The crew consisted of 50 merchant crew and 28 armed military personnel. By the time the second torpedo struck, all surviving crew had abandoned the ship in lifeboats with the exception of the Commander of the Armed Guard Unit and three enlisted men at the stern gun, who joined the crew after the second explosion. The crew spotted the U-boat before the firing of the second torpedo when it surfaced and moved toward the ship and lifeboats. Two rounds were fired from the submarine's deck gun at a 5 s interval directly after the release of the second torpedo, but it is not known if that was directed at the sinking tanker or the survivors (Burch, 1943: 4). Following the second torpedo, the submarine struck the stern of the No. 1 lifeboat, which raised it up from the water and knocked several survivors out, one of whom landed on the submarine before falling into the water. The submarine reportedly stayed in the area for 15–20 min but did not open fire on or attempt to capture any survivors. Survivors described the submarine based on their quick glimpses in the dark and a sketch was produced based on their observations (Fig. 9.2). A total of three crew were killed in the attack likely from the first explosion: third Assistant Engineer, Robert T. Winslow; Fireman, James B. Mitchell; and Oiler, Frank B. Robuck, all of whom were on duty in the engine room at the time of the attack (Moore, 1983). The 75 survivors were picked up by SC 1049 and brought to

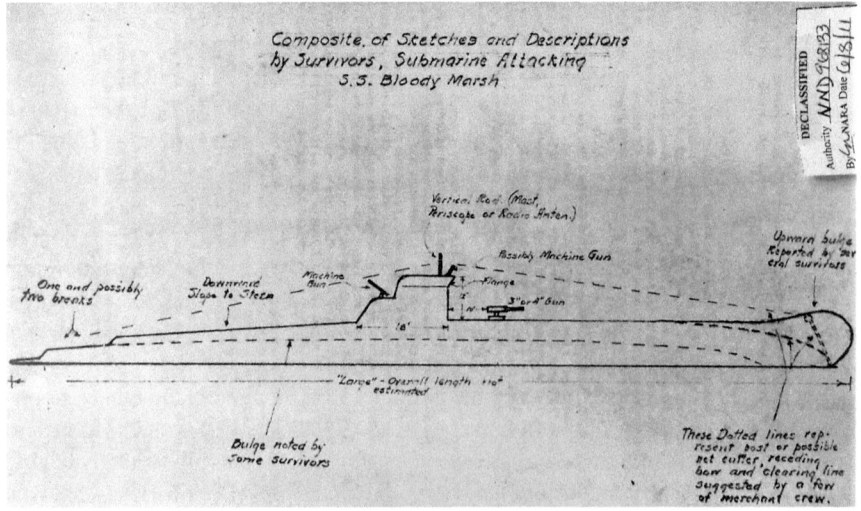

Fig. 9.2 Sketch of U-66 based on descriptions of survivors (U.S. National Archives)

Charleston, South Carolina the following morning. This attack is compared to U-66's sinking of SS *Esso Gettysburg* the month before off the east coast of Florida, a tanker that was traveling the same route as *Bloody Marsh*.

9.3   Initial Expedition

In 2018, NOAA Ship *Okeanos Explorer* published its planned mapping expedition location off the southeastern US and the first author submitted *Bloody Marsh* as a potential mapping target. Previously, the Kongsberg EM302 multibeam sonar was run over the wrecks of *Gulfoil* and *Gulfpenn* in the Gulf of Mexico from exploration vessel (E/V) *Nautilus*. Each wreck is at about 500 m depth and this was a test of the detection capability of the hull-mounted sonar over large oil tanker wrecks; they were visible but only as small bumps on the landscape. However, this indicated that the wreck of *Bloody Marsh* could be detected with hull mounted sonar. Multibeam expert Gary Fabian reviewed the data collected from *Okeanos Explorer* and found a target approximately the right size for the tanker, however when the ROV dived on it, it was determined to be a natural rock outcrop (Brennan, 2019).

Following this result, the geoscience company CGG reached out to maritime archaeologists at SEARCH through a mutual colleague at the Bureau of Ocean Energy Management (BOEM). CGG had previously licensed BOEM with an offshore natural oil seepage study across the US East Coast, which included satellite-detected oil slicks around the documented site of the *Bloody Marsh* recorded over several decades. It was hoped this specific oil slick dataset could help refine the search area for future explorations for the wreck.

9.4   Satellite Detection of Sea Surface Oil Slicks

The detection of oil slicks from sunken shipwrecks, and the usage of those oil slicks in the subsequent locating of a vessel is not in of itself a new technique. Following the loss of the bulk carrier MV *Derbyshire* in 1980, oil slicks were reported by patrol aircraft of the Japanese Maritime Safety Agency a week after the vessel's disappearance. These oil slick records, paired with ocean current models, allowed the successful discovery of the wreck location with side-scan sonar and an ROV dive in 1994 (Mearns, 1995).

The detection of the presence of oil slicks by satellite imagery is also not a new occurrence. Indeed, CGG were the first to extensively test new satellite systems in the early 1990s, partnering with the British National Space Centre (BNSC) and several oil companies at the time to test theories of oil slick detection. Since then, many new satellites have been launched, adding to the archive of satellite imagery capable of recording oil slicks. CGG has utilised this dataset to build a global database of oil slicks around the world.

The original BOEM natural oil seepage mapping study provided by CGG follows oil slick mapping and classification techniques first developed to document spatially repeating oil slicks across a satellite imagery stack over multiple imaging dates (Press & Lawrence, 1995). By observing a large imagery stack over a location across multiple dates, features can be observed that may be infrequent/ephemeral, whilst allowing for false positive features that may be misidentified on single imaging occurrences (e.g., oil slicks sourced from shipping) to be removed.

In the region of the proposed *Bloody Marsh* site, the study utilised 76 satellite images recorded from 1992 to 2019, of which 16 documented notable oil slicks (see Table 9.1 and Fig. 9.3). This imagery, a mixture of Synthetic Aperture Radar (SAR)

Table 9.1 Slicks observed across satellite imagery

Imagery date	Satellite	Imagery	Length of slick (m)	Origin point	Min Vol (m3)
1992-08-31	ERS-1	SAR	44,683	−79.040486, 31.464449	0.24
1997-03-27	ERS-2	SAR	9720	−79.049061, 31.458824	0.09
1997-06-24	RADARSAT-1	SAR	23,315	−79.023875, 31.454477	0.19
2008-06-12	ERS-2	SAR	50,812	−79.048919, 31.460945	0.42
2008-12-04	ERS-2	SAR	108,338	−79.048285, 31.402262	1.14
2010-01-28	ERS-2	SAR	85,662	−79.041723, 31.461843	0.8
2010-02-24	ALOS-1	SAR	5361	−78.780485, 31.555644	0.05
2013-07-26	LANDSAT 8	Optical	7176	−79.034845, 31.463299	0.02
2014-04-24	LANDSAT 8	Optical	13,835	−79.023249, 31.451961	0.04
2014-06-27	LANDSAT 8	Optical	28,083	−79.049815, 31.453468	0.18
2017-07-05	LANDSAT 8	Optical	6425	−79.057089, 31.454249	0.02
2019-01-17	SENTINEL-1A	SAR	28,822	−79.042821, 31.461263	0.47
2019-03-30	SENTINEL-1A	SAR	18,156	−79.045381, 31.455681	0.23
2019-04-22	LANDSAT 8	Optical	19,245	−79.048259, 31.450660	0.14
2019-05-24	LANDSAT 8	Optical	See Note	See Note	See Note
2019-08-09	SENTINEL-1A	SAR	12,773	−79.046534, 31.462054	0.12

Note: Two parallel slicks with potential affiliation to the Bloody Marsh were observed on the Landsat-8 image acquired on the 24th May 2019 and shown in Fig. 9.3

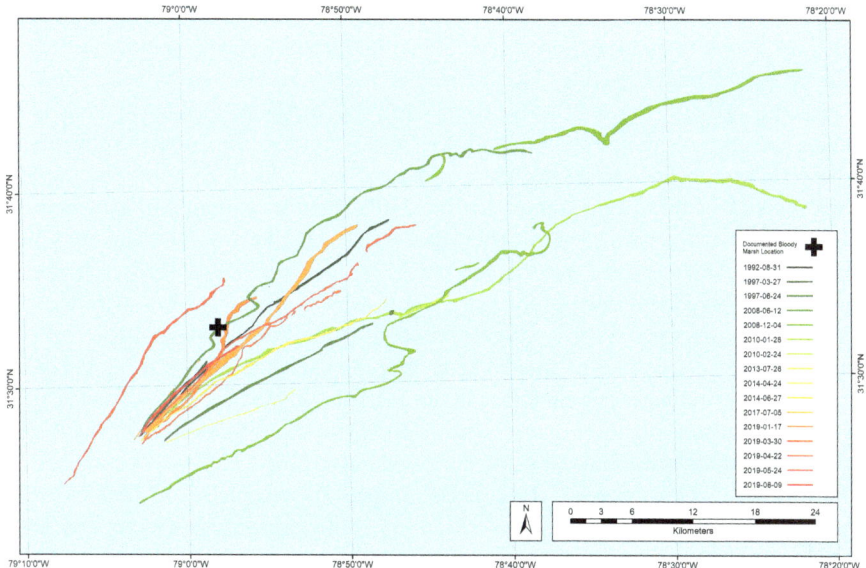

Fig. 9.3 Oil slicks mapped across 16 satellite images from 1992 to 2019, compared to the (imprecise) historically documented location of the *Bloody Marsh* wreck site. Whilst the majority of slicks originate from the southwest within a relatively constrained region with a radius of ~1.5 km, two notable outlier slicks are documented approximately 10 km away from this repeat location (imaged on 4th December 2008 and 24th May 2019). These outliers may reflect significantly different currents within the water column at the time of imaging or additional seabed oil sources

and multispectral, was sourced and processed from the Open Access archives of the European Space Agency (ERS and Sentinel-1), Canadian Space Agency (Radarsat-1), Japanese Space Agency (ALOS-1) and the United States Geological Survey (Landsat-8).

9.5   Oil Slick Evidence Towards *Bloody Marsh* Wreck Location

While only 20% of images used in the BOEM study documented oil slicks in the region, the reoccurrence of slicks across 27 years was highly indicative of a persistent fixed seafloor origin of the surface oil (see Fig. 9.3). Recent studies of other fixed point oil release sites around the world, at varying depths, carried out by CGG and others, resulted in the conclusion that these slicks showed strong spatial correlation to a single point source.

Typically, sea surface oil slicks fed from seabed sources (natural or anthropogenic) tend to develop curved 'corkscrew' morphologies as the overlying sea surface rotates due to near-inertial oscillation drift. This is due to most satellite scenes that observe oil slicks being acquired at low wind speeds, due to the requirement of relatively calm conditions for the slicks to form. However, the slicks observed in the *Bloody Marsh* region were observed to drift in the strong Gulf Stream currents off the coast of South Carolina. These currents, averaging 10.3 km/h, act to drift the surfacing oil slicks rapidly away from surfacing location, leading to long linear slicks ranging from approximately 5–110 km. Crucially, the surfacing location of these slicks (origination point) where observed in the same location and the long drifting slicks showed no sign of fragmentation before evaporation. This strongly suggested that the release of oil from the wreck was likely continuous, and those images in CGG's archive that did not record an oil slick where likely to be due to other mitigating met-ocean conditions. When utilising the multispectral data in particular, that images oil directly, thicker components of the oil slick could be observed at the southwestern end of the slick, in keeping with the theory of a subsurface release, before thinning and drifting in the currents (Fig. 9.4).

Importantly, the site of the recurring oil slicks mapped by CGG was 12 km southwest of the historically documented location of SS *Bloody Marsh* and previous surveys. Following on from these observations, 3D models of oil plumes in the water column were created based on the prevailing currents in the region. These allowed backtracking of the oil source from the sea surface slick observation locations towards a clear target area on the seabed.

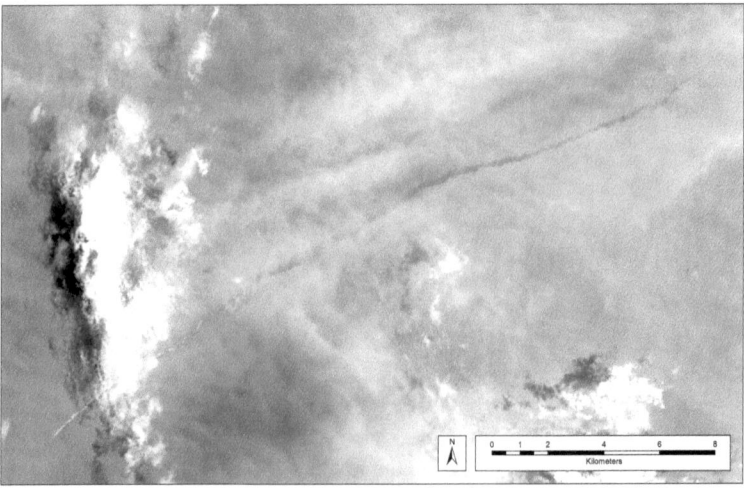

Fig. 9.4 Contrast enhanced Landsat-8 image of a sea surface oil slick (linear feature running southwest to northeast) acquired on 27th June 2014. The origin point of the oil slick can be determined as the southwestern end by the presence of thicker oil (silver/metallic appearance) which disperses and thins out into a sheen as currents move the oil slick away from the source point. Landsat-8 image courtesy of the U.S. Geological Survey

9.6 Discovery

An additional survey request was made to NOAA, and a target was detected only a few miles from the downstream limit of the oil slicks. The target measured shorter than the length of the tanker, and was a single target, where the reports of *Bloody Marsh*'s sinking have it breaking in half at the surface. However, the target was located in a relatively flat area of seabed two miles directly downstream of the oil slick origins on the surface, so was a promising target.

On October 28, 2021, ROVs *Deep Discoverer* and *Seirios* were launched from *Okeanos Explorer* to locate and identify the multibeam sonar target. Maritime archaeologists participated in the dive in real-time through telepresence capability on board the ship through NOAA's Ocean Explorer website and directed the dive remotely (Brennan et al., 2018). The ROV approached the target at 465 m depth and first came upon a debris field that was the remains of the destroyed stern and engine room. Immediately adjacent to this was the prominent hulk of the oil tanker completely inverted on the seabed (Fig. 9.5). A small amount of hull remained from the engine room, and a bulkhead between that space and the aft tanks appeared intact. Corroded holes were visible in the hull, but only those over machinery spaces. Like the *Coimbra* wreck, corrosion was more visible in areas of hull that had seawater on both sides; areas of hull that appeared to contain intact tanks had less corrosion (Brennan et al., 2023). Moving southeast forward along the wreck, the hull plate joinery indicated welded seams, which, along with the location, helped to positively identify the wreck as *Bloody Marsh*.

The main hulk of the wreck had few visible features, as the ship was completely turtled and the superstructure and decks buried in sediment. Biologists viewing the stream live commented that the wreck was colonised by what appeared to be a single species of gorgonian coral all of a similar size, indicating a single colonisation

Fig. 9.5 ROV image of the broken stern end of *Bloody Marsh* at the joinery between the engine room and cargo tanks (NOAA Ocean Exploration)

event. The ROV documented the starboard side of the wreck, but was unable to move around to the east-facing port side due to a strong current, therefore if any superstructure remains or debris lie on the seabed on that side, it was unable to be seen during this dive. When the ROV reached the forward end of the wreck, the measurement of the hull section indicated approximately 300 feet of hull. Subtracting the destroyed engine room, this would suggest that another 100 feet or so of hull remains. However, the bow section was not visible beyond the wreck and the current prevented further exploration in the area. The break in the hull, based on this measurement, appears to be at Tank #3, which is consistent with historic accounts. Therefore, the bow section could contain intact tanks #1 and 2 as well the as Deep and Fore Peak tanks forward. Future exploration in this area would be needed to locate and document this section.

9.7   Discussion and Conclusion

We had initially suspected *Bloody Marsh* could be a greater pollution risk than the PPW assessment projected due to its large cargo and the incident of its sinking. Had the vessel settled upright on the seabed, as the PPW report suggests is likely for wrecks in deep water, oil cargo would have escaped through the vents over time (NOAA, 2013a, b). This was the case for *Coast Trader* off the northwest Pacific coast (Delgado et al., 2018), as well as tanker wrecks in the Gulf of Mexico including *Virginia*, *GulfOil* and *GulfPenn* (Church & Warren, 2008), all of which sank intact in one piece. However, other oil tanker wrecks, such as *Coimbra* and *Munger T. Ball,* illustrate that ships sunk by torpedoes that break into multiple sections do not settle evenly on the seabed (Brennan et al., 2023). While those wrecks lie in shallower water, this proves to be the case for *Bloody Marsh* as well, and potentially indicates that other tankers sunk in deep water that remain undiscovered should also be considered high pollution potential as they may not have righted as they sank, especially if broken. This case study also serves as a proof of concept for utilising surface oil slicks as indicators of leaking oil tankers if oceanographic modeling can assist in back-calculating the wrecks' locations. As many of the PPW wrecks that remain to be discovered are in deeper water, this presents a promising method for locating other sites, especially those with cargos that are in danger of releasing.

The discovery of *Bloody Marsh* illustrates the complexity of locating shipwrecks in deep water. The size of the tanker and the depth at which it sank allowed for multibeam sonar to detect it once the satellite data redirected the search to a new area. This work shows the importance of technological advancements and data review to detect leaking wrecks and its potential to locate undiscovered sites. Additional survey in the area of *Bloody Marsh* is, however, still needed, as the 100-foot bow section remains missing. The second parallel slick identified in the satellite data could be this section, which is large enough to have some of the cargo tanks intact and still containing cargo. Further survey may locate this section potentially downstream of that second slick. The overturned hull of the tanker appeared stable, suggesting large amounts of oil could remain inside, and this site could be one to be considered for ROV-based mitigation in the near future.

Acknowledgements We would like to thank Pam Orlando and Frank Cantelas with NOAA for providing research materials from the National Archives, Kasey Cantwell, Matt Dornback and Derek Sowers from NOAA Ocean Exploration for coordinating the mapping and ROV dives, and Gary Fabian and Matt Horn for data review and advice.

References

Brennan, M. (2019). Search for SS *Bloody Marsh*. NOAA Ocean Explorer https://oceanexplorer. noaa.gov/okeanos/explorations/ex1903/logs/june29/june29.html. Accessed 1 Oct 2022.

Brennan, M. L., Cantelas, F., Elliott, K., Delgado, J. P., Bell, K. L. C., Coleman, D., Fundis, A., Irion, J., Van Tilburg, H. K., & Ballard, R. D. (2018). Telepresence-enabled maritime archaeological exploration in the deep. *Journal of Maritime Archaeology, 13*, 97–121.

Brennan, M. L., Delgado, J. P., Jozsef, A., Marx, D. E., & Bierwagen, M. (2023). Site formation processes and pollution risk mitigation of World War II oil tanker shipwrecks: *Coimbra* and *Munger T.* Ball. *Journal of Maritime Archaeology, 18*, 321–335.

Burch, H. A. (1943). SS 'BLOODY MARSH', Report of sinking by torpedoes. Division of Naval Intelligence, Counter Intelligence Branch. Tenth Fleet ASW Analysis & Stat Section, Series XIII, Reports and Analysis of U.S. and Allied Merchant Shipping Losses 1941–1945, Box 215. National Archives NND968133.

Church, R. A., & Warren, D. J. (2008). The 2004 deepwrecks project: Analysis of World War II Era shipwrecks in the Gulf of Mexico. *International Journal of Nautical Archaeology, 12*, 82–102.

Delgado, J. P., Cantelas, F., Symons, L. C., Brennan, M. L., Sanders, R., Reger, E., Bergondo, D., Johnson, D. L., Marc, J., Schwemmer, R. V., Edgar, L., & MacLeod, D. (2018). Telepresence-enabled archaeological survey and identification of SS Coast Trader, Straits of Juan de Fuca, British Colombia, Canada. *Deep-Sea Research Part II, 150*, 22–29.

Liddell, A., & Skelhorn, M. (2012). RFA Darkdale Survey Report, Salvage and Marine Operations, 173 pp.

Mearns, D. (1995). *Search for the bulk carrier Derbyshire: unlocking the mystery of bulk carrier shipping disasters*. Accessed 15 May 2023. https://www.researchgate.net/publication/254549249_Search_For_The_Bulk_Carrier_Derbyshire_Unlocking_The_Mystery_Of_Bulk_Carrier_Shipping_Disasters

Moore, C. A. R. (1983). *A careless word... A needless sinking*. American Merchant Marine Museum. Knowlton & McLeary.

NOAA. (2013a). *Risk assessment for potentially polluting wrecks in U.S. Waters*. https://nmssanctuaries.blob.core.windows.net/sanctuaries-prod/media/archive/protect/ppw/pdfs/2013_potentiallypollutingwrecks.pdf. Accessed 6 Apr 2022.

NOAA. (2013b). *Screening level risk assessment package: Bloody Marsh*. https://nmssanctuaries.blob.core.windows.net/sanctuaries-prod/media/archive/protect/ppw/pdfs/bloody_marsh.pdf. Accessed 6 Apr 2022.

Press, N., & Lawrence, G. (1995). Offshore basin screening from ERS satellite. In *Proceedings of the second ERS applications workshop*, London, UK, 6-8 December 1995. Accessed 15 May 2023. https://adsabs.harvard.edu/pdf/1996ESASP.383..129P.

Spyrou, A. G. (2006). *From T-2 to supertanker*. iUniverse, Inc..

Open Access   This chapter is licensed under the terms of the Creative Commons Attribution 4.0 International License (http://creativecommons.org/licenses/by/4.0/), which permits use, sharing, adaptation, distribution and reproduction in any medium or format, as long as you give appropriate credit to the original author(s) and the source, provide a link to the Creative Commons license and indicate if changes were made.

The images or other third party material in this chapter are included in the chapter's Creative Commons license, unless indicated otherwise in a credit line to the material. If material is not included in the chapter's Creative Commons license and your intended use is not permitted by statutory regulation or exceeds the permitted use, you will need to obtain permission directly from the copyright holder.

Chapter 10
Searching for a Lost PPW: SS *William Rockefeller*

John Detlie

10.1 Introduction

Possibly the most unfortunate legacy of the Battle of the Atlantic is the great number of wrecks lying in coastal and marine waters that still contain oil, unexploded ordnance, and other pollution or safety hazards. This has become an increasingly pressing issue in recent years, as these wrecks are now entering their eighth decade underwater and many are experiencing significant structural degradation due to natural corrosion and anthropogenic causes. While some of these potentially polluting wrecks have been located and steps taken to mitigate their pollution risk, many more remain missing, particularly those which sank in deep water (NOAA, 2013a, b; Symons et al., 2014; Hoyt et al., 2021; Brennan, Chap. 1, this volume).

This chapter discusses a potential method for locating these missing wrecks, using as a case study the oil tanker *William Rockefeller*. *Rockefeller* was a twin-screw, steel-hulled oil tanker, owned and operated by the Standard Oil Company of New Jersey. It was 554 feet long with a 75-foot beam and a gross tonnage of 14,054 tons. Its cargo capacity was 22,390 deadweight tons, or approximately 146,745 barrels of oil, carried in eight tanks amidships divided by 13 oil-tight bulkheads. When launched in 1921, *Rockefeller* and its sibling ship *John D. Archbold* were the largest oil tankers in the world. *Rockefeller* operated in the Gulf and Pacific oil trades for most of its career, though with the advent of World War II it began taking on cargo from international ports (Newport News Shipping and Dry Dock Company, 1920:9–20; Standard Oil, 1946:320–321).

J. Detlie (✉)
East Carolina University, Greenville, NC, USA
e-mail: detliej20@students.ecu.edu

© The Author(s) 2024   129
M. L. Brennan (ed.), *Threats to Our Ocean Heritage: Potentially Polluting Wrecks*, SpringerBriefs in Underwater Archaeology,
https://doi.org/10.1007/978-3-031-57960-8_10

On 28 June 1942, *Rockefeller* was sailing past Cape Hatteras en route from Aruba to New York with a cargo of heavy fuel oil when it was ambushed by the German submarine *U-701*. A single torpedo caught *Rockefeller* amidships, breaching one of its storage tanks, spraying oil across the deck, and setting the ship on fire. After its crew abandoned ship, *Rockefeller* was left to drift with the currents and wind before finally going down sometime later. There are several versions of *Rockefeller*'s sinking; contemporary accounts indicate that it either sank on its own or after *U-701* hit it with a second torpedo almost 12 h after the attack, while some secondary sources record a claim that it was sunk the following morning by the Coast Guard as a hazard to navigation (Fig. 10.1) (Degen, 1942; Standard Oil, 1946:321–323; Hoyt et al., 2021:7–278).

When it went down, *Rockefeller* was carrying 136,647 barrels (5,739,174 gallons or 21,725,136 liters) of heavy fuel oil, excluding its bunkered fuel. Some of this oil was spilled or burned off by *U-701*'s torpedoes, but it is probable that a significant

Fig. 10.1 Attack locations and possible sinking locations derived from various sources on *Rockefeller*'s loss. From bottom to top: the point at which *Rockefeller*'s known course diverged from prescribed convoy routes; attack location as given by uboat.net; attack location as given in the Office of the Chief of Naval Operations' (OCNO) supplemental statement to the summary of events given by *Rockefeller*'s crew; sinking location as given by uboat.net; attack location as given by *Rockefeller*'s master William Stewart; sinking location based on OCNO's coordinates, given a sinking time just before midnight of 29 June; sinking location based on Captain Stewart's account, given the above sinking time; sinking location based on OCNO's statement given an approximate sinking time in the early morning of 29 June; sinking location based on Captain Stewart's account, given the ship sinking on the morning of 29 June. (OCNO, 1942b, c; uboat.net 2021; map drawn by the author)

amount remains in the wreck (OCNO, 1942a, c; Standard Oil, 1946:321; NOAA, 2013b:13). Because of this, it is one of the 87 wrecks listed in the RULET database given the potential it poses for severe pollution of the Eastern seaboard. Since the wreck has not yet been located and its condition is unknown, NOAA's risk assessment screening package has recommended conducting surveys of opportunity to locate the vessel, determine its condition, and judge whether it is an imminent hazard (NOAA, 2013b:16–22, 38–39; Symons et al., 2014).

Since *Rockefeller* sank after drifting for at least 11 h, possibly more, and the only claimed eyewitness to its sinking did not give coordinates for its last resting place in his narrative of events, establishing a search area for the wreck has proven problematic (see Fig. 10.2). Some estimates have suggested that a search would encompass up to 750 square nautical miles (NOAA, 2011, 2013b:6). As this area would require significant outlay of time and effort to fully cover, this chapter discusses a means by which this potential search box may be shrunk to a more manageable size using Bayesian search theory, a method that has been tested and proven in multiple maritime searches. If successful in locating *Rockefeller*, this methodology may be useful in the creation of search models for other potentially polluting shipwrecks lost under similar circumstances.

10.2 Why Bayesian Search?

A shipwreck is typically a high-stress event involving imminent risk to life and limb, with accompanying repercussions for the thought processes, emotions, and memories of those involved (Gibbs, 2002:72–76). Even if someone thinks to record a sinking ship's last coordinates in the heat of the moment, it is very possible that these coordinates will be mistaken, or later misremembered. One has only to look at the records of past wreck search expeditions to see this problem in action: numbers are transposed, latitudes and longitudes are scrambled, and in some cases the coordinates given are simply wrong. Possibly the most notorious example of this phenomenon is that of the RMS *Titanic*. As the famed liner was sinking, wireless operators Harold Bride and Jack Phillips repeatedly transmitted distress signals containing what they believed to be an accurate position for the stricken ship. Subsequent search expeditions took this position data at face value and based their efforts on it, only to come away empty-handed. It was not until Dr. Robert Ballard reevaluated the problem, concluded that the positional data must be in error, and expanded the search area that *Titanic* was relocated in 1985, 13 nautical miles from the position given by Bride and Phillips (Ballard, 1987:23, 26, 66, 83).

The problem for the archaeologist or historian is how to sift through these disparate or sparse sets of information in such a way as to maximise their chances of completing a successful search. Then comes the second issue, which is equally important: selecting a theoretical framework/model into which this data can be fed to generate a search plan. All this is done with an eye toward optimising the search parameters so as to produce the highest chance of success with the least expenditure of money, time, and effort. Dr. Lawrence Stone refers to this as the basic problem of optimal search (Stone, 1975:32). There have been a variety of methods proposed for the solution of this problem, but Bayesian search theory offers a method that accounts for all possible scenarios while still allowing researchers and archaeologists to focus their efforts on a smaller, high-priority search area. It has been successfully employed in multiple maritime search efforts, including the recovery of an H-bomb lost off the coast of Spain in 1966 and locating the wrecks of USS *Scorpion*, SS *Central America*, and Air France Flight 447 (Richardson & Stone, 1971:141–144; Stone, 1992:42–53; Sontag et al., 1998:58–60, 104–106; Craven, 2001:167–170, 173–174, 213; Frost & Stone, 2001:3–4; Stone, 2011:21, 23; Stone et al., 2014:72–80).

Bayesian search theory is based on the statistical models of eighteenth-century English statistician and philosopher Thomas Bayes (Bayes & Price, 1763). Put simply, Bayesian statistics operate on a definition of probability wherein 'probability' expresses a degree of belief in the occurrence of a given event. This belief may be based on prior knowledge of the event, whether derived from personal experience or historical records, or it may be based on one's own theories or beliefs about said event. The probabilities are expressed as part of an equation which is used for a given purpose, in this case searching for a lost shipwreck.

Bayesian theory as applied to maritime search is a relatively straightforward concept. The researcher must form as many reasonable hypotheses as possible about what may have happened to their target object, based either on hard data or their own beliefs. These hypotheses are used to generate a probability density function for the object's location, which calculates the probability of a random variable (in this case, the location of a submerged object) falling within a particular range of values (Stone, 2011:23).

After this, a second function is constructed which expresses the likelihood that the target object will be found at a given location, *if it is really in that location*. This information is used to generate a probability map, which gives the probability of finding the target object in a given location for all possible locations within the projected search area. After this, a search grid and path is devised that covers the entire area from highest to lowest areas of probability. During the search, the probabilities must be continually revised according to the findings. For example, if the object is believed to have fragmented before sinking, and fragments are not found in the areas where they are most likely to be according to the map, then the fragmentation hypothesis becomes less probable and should be revised or discarded accordingly (Stone, 2011:23–24).

As new data is gathered, it is fed into the model to update the probabilities contained therein. The model, of course, is only as good as the data that is used to make it, so someone who employs Bayesian search theory must be sure that they are collecting and using data that is as accurate as possible. Given the problems mentioned at the beginning of this section, this is not always an easy task. Such proved to be the case when applying the principles of Bayesian search to the problem of finding *William Rockefeller*.

10.3 Searching for *Rockefeller*

In the case of *William Rockefeller*, there existed several eyewitness accounts of the attack on the tanker, including its master William Stewart and Kapitänleutnant Horst Degen, the captain of *U-701*, who was captured and interrogated after his submarine was sunk in the vicinity of Cape Hatteras on 7 July 1942. These accounts form the basis of most of the research done on the tanker's loss; while they are in general agreement as to *Rockefeller*'s last hours afloat, there were some discrepancies that had to be accounted for. Most of these discrepancies could be put down to the stress of the sinkings and the unreliability of human memory, though there were indications of deliberate obfuscation or selective recall, especially when comparing the initial accounts of the attack to those set down years or decades later (Degen, 1942; OCNO, 1942a, c; Standard Oil, 1946; Offley, 2014). The accounts were used to construct a series of scenarios for *Rockefeller*'s loss, with the aim of covering all the possibilities as to the tanker's fate. They may be summarised as follows:

- Scenario 1: *Rockefeller* sinking on its own after drifting and burning for just over 11 h. This is based on the reports of the US Navy and Coast Guard.
- Scenario 2: *Rockefeller* being sunk by a second torpedo from *U-701*, after ~12 h adrift. This is based on the interrogation of KptLt Degen and his unpublished postwar account.
- Scenario 3: *Rockefeller* being scuttled by Coast Guard aircraft on the morning of 29 June. This is based on the accounts of several secondary sources which record a Coast Guard report to this effect.

As is obvious, the scenarios cannot all be correct, so what weight is to be given to each of them? This is a task requiring subjective probability analysis; in plain language, this analysis involves a person looking at all the information they have gathered and deciding what they believe to be the most reliable. This is a part of the Bayesian method that is meant to quantify the otherwise unquantifiable human factors of intuition, gut feelings, and hunches. This is not to say that the process is unscientific, for it requires the person or persons conducting the analysis to carefully study the available data and evidence when making their decisions, but it contains an element of uncertainty since it is based on subjective opinions (Stone, 1992:45–46; Sontag et al., 1998:59, 104–105; Craven, 2001:167–168). In this case, the author decided that Scenario 2 was the most likely, since it tracked closely with what is generally accepted about *Rockefeller*'s last hours and Degen provided a persuasive account of his claimed second attack on the submarine (USONI, 1942; Offley, 2014). Scenario 1 overlaps in most particulars with Scenario 2; it may even be the case that Degen's second attack on *Rockefeller* was seen and processed as the tanker sinking on its own. These two scenarios were therefore weighted as being equally likely. The third scenario is not well-supported by the primary sources and appears mainly in secondary sources; it was therefore rated as the least likely of the three.

With the scenarios devised and weighted, a probability map was created using ArcGIS and SAROPS, a software package developed for the US Coast Guard that employs Bayesian mathematics and Markov chain Monte Carlo simulations to generate probability maps for maritime search-and-rescue efforts (United States Coast Guard, 2008; Kratzke et al., 2010:1–2).

Monte Carlo simulations are used to calculate the results of a scenario with numerous variables, expressing the outcome as a probability distribution showing the range of possible outcomes according to the likelihood of their occurrence. However, a standard Monte Carlo simulation is not suitable for more complex problems, since it relies on the assumption that the values for which it calculates results are independent of each other and may be drawn independently. The solution to this problem is to use a Markov chain, which calculates each successive variable based on the value of the last variable generated, creating a stochastic chain that allows for greater flexibility in generating the final probability density (Brownlee, 2019).

SAROPS uses Markov chain Monte Carlo simulations to generate a probability density function by sequentially generating random variables based on the data that is fed into the program. For a typical SAR scenario, this data is acquired in real time, which enables the production of highly accurate maps (Roylance, 2007:4D). In this case the simulation had to rely on historical data, which added a greater degree of uncertainty to the outcome. The two main variables were the currents and wind speed prevailing during the attack; a combination of interpretations and contemporary data had to be employed to reconstruct them.

The attack occurred within the Gulf Stream, meaning that *Rockefeller*'s drift after being abandoned would have been influenced by this relatively strong and predictable current. Its precise speed on 28 June 1942 is not known, but the Coast Guard casualty report for *Rockefeller* indicates the ship was drifting at 1.5 kt when abandoned (USCG, 1944b). Data obtained from a 1942 Coast and Geodetic Survey publication indicated that the Gulf Stream was then known to flow strongly northeastward during the summer months, peaking in strength during July. The monthly average of the Gulf Stream in June as measured by the Diamond Shoals lightship from 1919 to 1928 was 0.66 kt (Haight, 1942:24–25, 52). It is therefore probable that the Stream was pushing *Rockefeller* northeast at low speed. Similar examples of Gulf-influenced drift can be seen in the cases of the merchant ship *Papoose* and *U-701*. *Papoose* was torpedoed approximately 15 miles southwest of Cape Lookout, North Carolina, and drifted north with the Stream for 2 days before sinking off Oregon Inlet, and the survivors of *U-701* were carried northeast with the current for 49 h before being recovered by the Coast Guard (USCG, 1944a; USONI, 1942:1; Hickam, 1989:278–282; NOAA, 2022).

Wind speed is another variable for which assumptions and estimates had to be made, though in this case more solid data was available. *Rockefeller* was being escorted by a Coast Guard cutter that logged weather data at 4-h intervals, noting general conditions, barometric pressure, cloud cover, visibility, and wind speed. At noon, 16 min before *Rockefeller* was attacked, the cutter logged the following weather data: barometric pressure 3001, winds from the southwest at Beaufort

Force 1 (1–3 mph/1–3 kt), blue skies with scattered cumulonimbus clouds, visibility 7 miles (USCG, 1942:2). Given that *Rockefeller* was drifting at about 1.5 kt when it was abandoned, it may be presumed that the wind was very light at this time, possibly not more than 1 kt, therefore both wind and current would have been exerting force against the tanker's hull.

These estimates were added to a data package which also included events, times, *Rockefeller*'s speed (known or estimated), *Rockefeller*'s position, general weather conditions, and compass bearings. The package was then sent to the Coast Guard to be input into SAROPS. Senior Chief Petty Officer Ian Brown, an experienced SAROPS technician, was given the task of creating the SAROPS model. He consulted with oceanographer Dr. Cristina Forbes, who helped him format the data for SAROPS, calculated possible wind and current speeds, and proposed that he create three different scenarios based on possible weather conditions at the time. After this consultation, Chief Brown fed the resulting data into the SAROPS simulator. He explained to the author that the maximum vessel length that could be input into the program was 300 feet, since the program is tailored toward searching for lost individuals and small craft, not large vessels. As *Rockefeller* was 554 ft. in length, this meant that he had to split the simulated vessel into two parts and recombine the results later. He also input the estimated current, wind, and drift speeds from the historical data and his consultations with Dr. Forbes. All times in the program were recorded in Greenwich Mean Time (GMT) and were then adjusted into Eastern War Time (EWT) (Ian Brown 2022, pers. comm.).

To account for the uncertainty of wind speeds, Chief Brown ran multiple iterations of the scenario. For the first iteration, he used a wind speed of 0.87 kt and a current speed of 0.66 kt until time index 1700 GMT (1300 EWT), representing the likely point at which *Rockefeller* would have been in the middle of the Gulf Stream and experiencing its full strength, which he calculated at 1.5 kt. For the second iteration of the scenario, he used a wind speed of 1.74 kt and the same current estimates as in the first iteration, with *Rockefeller* entering the middle of the Gulf Stream at time index 1700 GMT/1300 EWT. For the third iteration, he again used the same current estimates and a wind speed of 2.6 kt. Each of these iterations produced a probability density cloud illustrating potential sinking locations for *Rockefeller* as based on the estimates fed into the program. For each stage of the scenarios, he placed a ring around the density cloud to indicate the general size of the area of potential drift, given in square nautical miles (Ian Brown 2022, pers. comm.).

These density clouds were added to an ArcGIS map containing the sinking locations from the historical records and a drift analysis created by the author to produce a final probability density map (see Fig. 10.2). A search grid was imposed over the area of highest density with cells measuring 16 square nautical miles each; this grid covers a total of 384 square nautical miles. The potential search area has thus been almost halved from NOAA's original estimate of 750 nautical miles. If sectioned into a grid of 4 by 4 nm segments, this gives 24 cells in which to search within this box (see Fig. 10.3). The question that follows is how to further prioritise the search area to maximise the chances of success. According to Bayesian methods, the

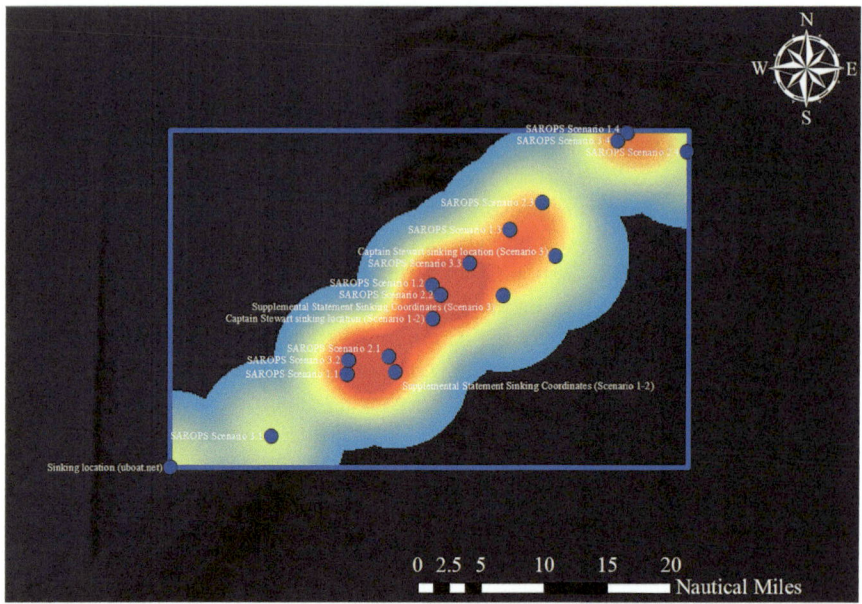

Fig. 10.2 Combined density cloud containing data points from the author's GIS work and the SAROPS simulations. (Map created by the author and Senior Chief Ian Brown)

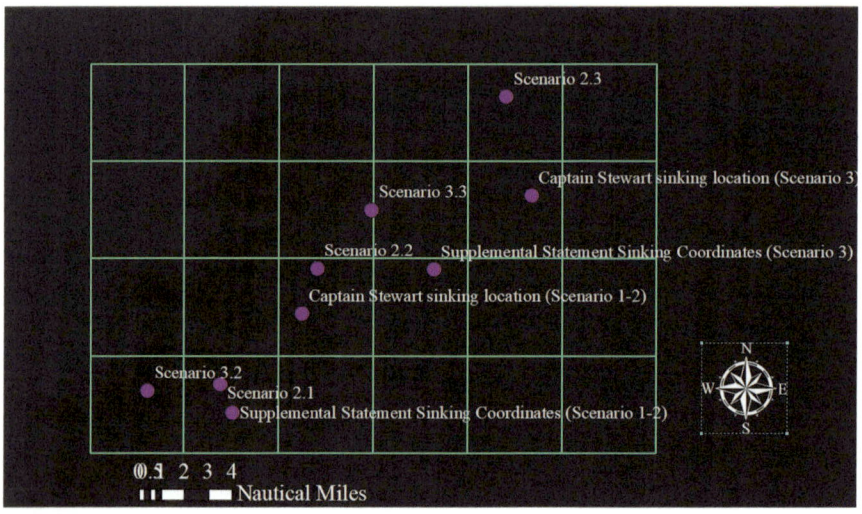

Fig. 10.3 Calculated sinking locations with priority search grid overlaid. (Map drawn by the author)

factors that must be considered here are the following: the *prior distribution* of information, the *subjective probabilities* of the occurrence of given events, and *posterior distribution* of data obtained from previous search efforts (Stone, 2011; Stone et al., 2014; Rossmo et al., 2019).

When speaking of maritime search, the *prior distribution* consists of any data collected prior to the current operation. As an example, we may consider the prior distribution of the 2011 search for Air France Flight 447's black box, which was conducted by Dr. Lawrence Stone using Bayesian methods. In that case, the prior distribution consisted of the last GPS ping received from Flight 447 before its disappearance, the estimated distance the plane might have traveled from its last known position based on the cessation of signal data, and the recovery locations of debris and victims from the wreck, which was used to plot a reverse-drift scenario to estimate the crash location (Stone et al., 2014:72–75).

In the context of the search for *Rockefeller*, the prior distribution consists of the historical and spatial data assembled regarding *Rockefeller*'s loss and the maps and models created from this data. The balance of the data suggests strongly that *Rockefeller* drifted slowly northeastward after being abandoned, probably not more than 17 nm and almost certainly no more than 25 nm. There is some margin of error here; *Rockefeller*'s given drift speed of 1.5 kt is only an approximation based on the Coast Guard casualty report, and it is possible that the winds may have shifted, pushing it in another direction. This factor was accounted for in the Coast Guard's SAROPS probability maps, which posited three different wind speeds and applied different current speeds as well throughout the scenario.

The next factor to be considered is the *probability of events occurring*. This is a subjective, not objective factor, since in this case the probability is based on what the person conducting the study knows or believes to be true. As stated previously, this study has accepted KptLt Degen's account of sinking *Rockefeller* as the most probable scenario with allowance for the fact that the tanker may also have sunk on its own around the same time, as stated in Scenario 1. The third scenario, in which *Rockefeller* was scuttled by the Coast Guard, is regarded as less probable and has been weighted accordingly. Each scenario was given a percentage, all of which added to 100: Scenario 1 = 45%, Scenario 2 = 45%, Scenario 3 = 10%. If a search is undertaken, these probability percentages can be fed into the basic Bayesian probability equation, which runs as follows:

$$P(A \mid B) = \frac{P(B|A)P(A)}{P(B)}$$

where P (A|B) is the conditional probability of event A given event B;
P (B|A) is the conditional probability of event B given event A;
P(A) is the probability of Event A;
P(B) is the probability of Event B (Rossmo et al., 2019:45).

In this case, there are three events to solve for, those being the scenarios described previously. Event A may be given as Scenario 1 (*Rockefeller* sinking on its own), Event B as Scenario 2 (Degen sinking *Rockefeller* with a second attack), and Event C as Scenario 3 (*Rockefeller* being scuttled by the Coast Guard). Some of these equations are zero-sum games. Degen's account and the scuttling scenario cannot both be true, thus the conditional probability of Event B given Event C would be zero, and vice versa. The only way to know for sure is to go and look. When a search operation is launched, this equation and its derivatives may be used to update the probabilities according to the search finds, with the results fed into the predictive model in ArcGIS to create an updated search map.

The *posterior distribution* of information is assembled from all data acquired during previous search efforts. In the case of AF Flight 447, the posterior distribution included various unsuccessful searches for the plane's wreckage in 2009 and 2010, which Dr. Stone and his colleagues factored into their probability maps (Stone et al., 2014:75–79). In the context of this study, there is no posterior distribution to consider, as there have been no searches conducted for *Rockefeller*'s remains. It may be the case, of course, that this study's conclusions are in error and a search based on them fails to locate *Rockefeller*'s wreck. In this case, a posterior distribution can be created from the data obtained in this search and used to refine the search area for the next expedition.

In the context of the initial search for *Rockefeller*'s wreck, the areas of greatest coordinate density are the four cells in the center of the grid and the two cells at its lower left edge. Focusing the search on these cells further reduces the priority search area from 384 square nautical miles to 96 sq. nm. Searching these grid cells is therefore the logical place to begin. Should the ship not be found in these areas, the remainder of the grid can be covered in descending level of priority, either in the same expedition or a follow-up search.

10.4   Conclusion

This chapter has presented a method for locating lost potentially polluting shipwrecks in American waters, using Bayesian search theory and computer modeling to calculate the potential drift of an abandoned ship prior to its sinking. The initial results have produced a more manageable search box that can be further subdivided into smaller search areas based on the time and resources available. Furthermore, these search areas can be easily covered by existing underwater search technologies such as side-scan sonar and autonomous underwater vehicles. Should this method prove successful in locating the wreck of *William Rockefeller*, it can be repeated for other missing potentially polluting wrecks, provided sufficient data is available.

Acknowledgments The author gratefully acknowledges the cooperation and contributions of Ole Varmer and the United States Coast Guard in this study.

References

Ballard, R. D. (1987). *The discovery of the Titanic*. Warner Books.

Bayes, T., & Price, R. (1763). An essay towards solving a problem in the doctrine of chances. By the late Rev. Mr. Bayes, F. R. S. Communicated by Mr. Price, in a letter to John Canton, A. M. F. R. S. *Philosophical Transactions of the Royal Society of London, 53*, 370–418.

Brownlee, J. (2019). *A gentle introduction to Markov Chain Monte Carlo for probability*. https://machinelearningmastery.com/markov-chain-monte-carlo-for-probability/. Accessed 21 Oct 2022.

Craven, J. P. (2001). *The Silent War: The Cold War Battle beneath the sea*. Simon and Schuster.

Degen, H. (1942). Kriegstagebuch of U-boat 'U-701', Third War Patrol, April 1– July 7, 1942 (reconstructed at BdU Headquarters from BdU KTB and message traffic from U-701).

Frost, J. R., & Stone, L. (2001). Review of search theory: Advances and applications to search and rescue decision support. Joint publication, Zosa & Company Ltd., Metron, Inc., and U.S. Coast Guard Research & Development Center. Prepared for U.S. Department of Transportation United States Coast Guard Office of Operations Washington, DC. Report No. CG-D-15-01.

Gibbs, M. (2002). Maritime archaeology and behaviour during crisis: The wreck of the VOC ship *Batavia* (1629). In R. Torrence & J. Grattan (Eds.), *Natural disasters and cultural change* (pp. 66–86). Routledge.

Haight, F. J. (1942). *Coastal currents along the Atlantic Coast of the United States: Special publication 230*. United States Department of Commerce, Coast and Geodetic Survey.

Hickam, H. (1989). *Torpedo junction: U-Boat War off America's East Coast, 1942*. Naval Institute Press.

Hoyt, J., Bright, J., Hoffman, W., Carrier, B., Marx, D., Richards, N., Sassorossi, W., Davis, K., Wagner, J., & McCord, J. (2021). Battle of the Atlantic: A catalog of shipwrecks off North Carolina's coast from the Second World War. Joint publication by US Dept. of the interior, Bureau of Ocean Energy Management, Sterling, VA and National Oceanic and Atmospheric Administration, Office of National Marine Sanctuaries, Newport News, VA. OCS Study BOEM 2021-076/NOAA Maritime Heritage Report Series Number 6.

Kratzke, T., Stone, L., & Frost, J. R. (2010). Search and rescue optimal planning system. In *Proceedings of the 13th international conference on information fusion* (pp. 1–8).

National Oceanic and Atmospheric Administration. (2011). *Battle of the Atlantic: Vessels of Interest*. https://sanctuaries.noaa.gov/missions/2011battleoftheatlantic/vessels.html. Accessed 14 Feb 2023.

National Oceanic and Atmospheric Administration. (2013a). Screening level risk assessment package: *Papoose*. National Oceanic and Atmospheric Administration, Office of National Marine Sanctuaries and Office of Response and Restoration, Washington, DC.

National Oceanic and Atmospheric Administration. (2013b). Screening level risk assessment package: *William Rockefeller*. National Oceanic and Atmospheric Administration, Office of National Marine Sanctuaries and Office of Response and Restoration, Washington, DC.

National Oceanic and Atmospheric Administration. (2022). Monitor national marine sanctuary shipwrecks: Papoose. https://monitor.noaa.gov/shipwrecks/papoose.html. Accessed 11 Sep 2022.

Newport News Shipbuilding and Dry Dock Company. (1920). Specifications for the construction of two steel twin-screw steamships: For carrying oil in bulk for the Standard Oil Co. of New Jersey. Newport News Shipbuilding and Drydock Company, Newport News, VA.

Office of the Chief of Naval Operations (OCNO). (1942a). Summary of statements by survivors of SS *William Rockefeller*, Standard Oil Company of New Jersey, enemy attack on Merchant Ships. Records of the Office of the Chief of Naval Operations, OP-16-B-5, July 10, 1942, National Archives and Records Administration, College Park, Maryland, Box 698, Records Group 38.

Office of the Chief of Naval Operations (OCNO). (1942b). Supplement to the summary of statements by survivors of SS 'WILLIAM ROCKEFELLER', U.S. tanker, 14,054 G.T. owned and operated by Standard Oil Co. of New Jersey. Records of the Office of the Chief of Naval Operations, OP-16-B-5, July 10, 1942, National Archives and Records Administration, College Park, Maryland, Box 698, Records Group 38.

Office of the Chief of Naval Operations (OCNO). (1942c). Affidavit of William R. Stewart, Master SS *William Rockefeller*, Torpedoed and Sunk, June 28, 1942. National Archives and Records Administration, College Park, Maryland, Records Group 38, Tenth Fleet ASW Analysis and Statistics Section, Series XIII, Report and Analyses of U.S. and Allied Merchant Shipping Losses, *Williams D. Burnham – YP-453*, Box 185.

Offley, E. (2014). *The burning shore: How Hitler's U-boats brought WWII to America*. Basic Books.

Richardson, H. R., & Stone, L. (1971). Operations analysis during the underwater search for *Scorpion*. *Naval Research Logistics Quarterly, 18*(2), 141–157.

Rossmo, D. K., Velarde, L., & Mahood, T. (2019). Optimizing wilderness search and rescue: A Bayesian GIS analysis. *Journal of Search and Rescue, 3*(2), 44–58.

Roylance, F. D. (2007). Lost at sea and rescued by modern technology. *Baltimore Sun* 6 April (96), 1D, 4D. Baltimore, MD.

Sontag, S., Drew, C., & Drew, A. L. (1998). *Blind man's bluff: The untold story of American submarine espionage*. PublicAffairs.

Standard Oil Company of New Jersey. (1946). Ships of the Esso Fleet in World War II. Standard Oil Company (New Jersey), New York.

Stone, L. D. (1975). *The theory of optimal search*. Academic.

Stone, L. D. (1992). Search for the SS *Central America*: Mathematical treasure hunting. *Interfaces, 22*(1), 32–54.

Stone, L. D. (2011). Operations research helps locate the underwater wreckage of air France flight AF 447. *Phalanx, 44*(4), 21–27.

Stone, L. D., Keller, C. M., Kratzke, T. M., & Strumpfer, J. (2014). Search for the wreckage of air France flight AF 447. *Statistical Science, 29*(1), 69–80.

Symons, L., Michel, J., Delgado, J., Reich, D., French McCay, D., Etkin, D. S., & Helton, D. (2014). The Remediation of Underwater Legacy Environmental Threats (RULET) risk assessment for potentially polluting shipwrecks in U.S. Waters. In *Proceedings of the 2014 International Oil Spill Conference*. Allen Press, Inc., Lawrence, KS, pp. 783–794.

United States Coast Guard. (1942). Deck Logs, 'USS *CG-470*'. National Archives and Records Administration, Records Group 26, Records of the US Coast Guard, Entries (NC-31) 159 D-I, Box 262. Washington, DC.

United States Coast Guard. (1944a). Report on U.S. Merchant Tanker War Action Casualty, *S/S Papoose*. War Casualty Section, Casualty Reports 1941 to 1946, Records of the United States Coast Guard, Entry 191, Box 5, Records Group 26, National Archives Building, Washington, DC.

United States Coast Guard. (1944b). Report on U.S. Merchant Tanker War Action Casualty, *S/S William Rockefeller*. War Casualty Section, Casualty Reports 1941 to 1946, Records of the United States Coast Guard, Entry 191, Box 5, Records Group 26, National Archives Building, Washington, DC.

United States Coast Guard. (2008). *Search and rescue optimal planning system information sheet*. Office of Search and Rescue, United States Coast Guard. https://www.dco.uscg.mil/Portals/9/CG-5R/SARfactsInfo/SAROPSInforSheet.pdf. Accessed 15 Nov 2022.

United States Navy Office of Naval Intelligence. (1942). Report of Interrogation of Survivors of *U-701*, Sunk by U.S. Army Attack Bomber No. 9-29-322, Unit 396 B.S. [Bombardment Squadron] on July 7, 1942. Post Mortems On Enemy Submarines – Serial No. 3. Office of Naval Intelligence, ONI 250-G, National Archives and Records Administration, College Park, Maryland, Records Group 38.

Open Access This chapter is licensed under the terms of the Creative Commons Attribution 4.0 International License (http://creativecommons.org/licenses/by/4.0/), which permits use, sharing, adaptation, distribution and reproduction in any medium or format, as long as you give appropriate credit to the original author(s) and the source, provide a link to the Creative Commons license and indicate if changes were made.

The images or other third party material in this chapter are included in the chapter's Creative Commons license, unless indicated otherwise in a credit line to the material. If material is not included in the chapter's Creative Commons license and your intended use is not permitted by statutory regulation or exceeds the permitted use, you will need to obtain permission directly from the copyright holder.

Chapter 11
Assessment Methodologies for Potentially Polluting Wrecks: The Need for a Common Approach

Mark Lawrence, Stuart Leather, and Simon Burnay

11.1 Introduction

There are believed to be more than 8500 Potentially Polluting Wrecks (PPW) lying on the seabed around the world, including many oil tankers, which potentially still contain very large quantities of oil. With many of these wrecks having been submerged for nearly 80 years (and sometimes more), the condition of their hulls is continuing to deteriorate, which combined with the increase in the frequency of severe weather events, means that the risk of significant pollution is increasing, particularly for wrecks in shallower waters.

The release of large quantities of oil from the catastrophic collapse of these wrecks is therefore an increasing risk. The resulting major pollution incidents would cause significant damage to the ocean environment, biodiversity and blue economies. The impact on the Global South coastal states, where there is a higher reliance on fishing communities and ocean economies, would be acute, as there is very limited capacity to deal with these events in those areas.

To understand the risk of pollution emanating from any wreck, the wreck must be assessed to determine its condition, the rate of deterioration, the likely pollutants remaining on board, the impact of any pollution and what methods might be feasible for preventing a release of pollutants into the ocean environment. From this assessment, viable management plans and intervention strategies for each wreck can be developed. The management and intervention of wrecks can then be prioritised according to the risk they present.

M. Lawrence (✉) · S. Leather (✉) · S. Burnay
Waves Group Ltd., London, UK
e-mail: m.lawrence@waves-group.co.uk; s.leather@waves-group.co.uk;
s.burnay@waves-group.co.uk

© The Author(s) 2024   143
M. L. Brennan (ed.), *Threats to Our Ocean Heritage: Potentially Polluting Wrecks*, SpringerBriefs in Underwater Archaeology,
https://doi.org/10.1007/978-3-031-57960-8_11

11.2   Current Status of Wreck Assessment Methods

A number of nations have developed methods for the risk assessment and management of PPW, including comparison between different wrecks in order to prioritise the need for intervention, according to the risk of the wreck releasing hydrocarbons and the potential environmental impact of a release.[1]

However, the methods developed to date are considered to be limited in certain areas and lack key components when evaluated against an international standard on risk management (e.g., ISO 31000: 2009). Few methods are considered to take into account the uncertain nature of many of these wrecks, or the risk of continuous low levels of oil release into the marine environment and there is little in the way of standardisation in terms of the methodology. The existing methods have tended to be national or regional only, and their deployment to actual wreck management and intervention projects has been sporadic. In our view, this is primarily due to budget constraints and the challenges of cross-border responsibility.

There is an absence of common standards and protocols for the assessment and prioritisation of PPW which can inhibit proactive risk management. This also inhibits funding for the management and intervention of PPW. Proactive management and intervention of PPW will reduce, and aims to eliminate, the risk of serious pollution incidents from PPW. This also significantly reduces costs; the cost of proactively removing pollutants from a wreck can be orders of magnitude less than dealing with an emergency pollution clean-up and response, let alone the environmental and socio-economic impacts of spills.

This chapter outlines the experience that we have gained at Waves Group in the assessment of PPW and illustrates how the technologies and methodologies used for assessment can be effective and can be standardised but are also readily adapted.

11.3   Our Experience in the Assessment of Potentially Polluting Wrecks

The assessment of PPW requires a range of technical disciplines, including a combination of high-resolution survey data acquisition, naval architecture, marine engineering, modern ship salvage and pollution expertise, together with maritime archaeology expertise to address the pollution risks and the cultural heritage aspects. We have integrated this combination of expertise, regularly deploying it on modern, commercial salvage and wreck removal projects. Removal of pollutants and managing pollution risks is at the heart of modern ship salvage and so the techniques used are directly applicable to PPW and therefore provide a proven, state-of-the-art capability to meet the PPW challenge. Applying adaptions of proven commercial

---

[1] Particular examples include VRAKA (Sweden), NOAA (USA), South Pacific Regional Environment Program (SPREP), SWERA (Finland).

technology, such as non-intrusive sampling for hydrocarbons, has enabled the development of methods that considerably reduce the risks of the destabilisation of the wreck and the unintended release of hydrocarbons into the marine environment.

We are working internationally with Governments and their Agencies, together with other stakeholders and experts in the PPW field, and have built experience across PPW projects in UK, US, Finnish and Pacific Island waters, and modern salvage projects worldwide, including many environmentally and socio-economically sensitive regions from the high Arctic to South Pacific islands. This includes both planned PPW assessments and emergency response interventions. Our experiences to date have enabled us to consider what is 'best practice' for the management of PPW and the technologies and methodologies used for their assessment, and also highlight the gaps and future needs for proactive wreck management and intervention. Here, we summarise some of the aspects of the methodologies that we have deployed and experienced.

The approach to the assessment of PPW is driven by the need to achieve the best possible certainty, whilst having the lowest environmental risk, and to ensure that the assessment can be conducted cost-effectively. The stakeholders to any PPW require a clear and accurate understanding of the risk presented by the wreck to justify the intervention strategy. The decisions required to be made are significant and clear, accurate data underpins this.

For a planned (i.e., proactive) wreck assessment, a comprehensive desktop scoping exercise is undertaken before any fieldwork. This includes detailed archival research to source all available technical data and documentation for the wreck (e.g., plans and drawings), details of the voyage and events surrounding the sinking, location specific information, including ocean data, environmentally sensitive areas near the wreck and geological information on the nature of the surrounding seabed. Any previously available data that may have been acquired is also integrated to this assessment, which is then used to assess the nature of the assets and equipment best suited to undertake the offshore elements of the assessment, safely and cost effectively.

To assess the potential for oil (or other pollutants) to remain in a wreck, it is critical to be able to identify the internal compartments of the wreck and to assess those that would be holding the pollutants. Using a combination of the archival research and state-of-the-art 3D digital models derived from vessel plans and lines drawings, with expert Naval Architect input, the internal arrangement and compartments of the vessel can be modeled and used to identify the potential sources of pollutants. Figure 11.1 shows the digital model of SS *Derbent* assessed on behalf of the UK MOD (see Hill et al., Chap. 6, this volume). The model has been derived from ships plans of the vessel class of tanker. The model clearly defines the internal structure of the vessel and the location of the cargo holds.

Fig. 11.1 Digital model of SS *Derbent* showing internal cargo tank structure

Fig. 11.2 High-resolution multibeam bathymetry data of the *Derbent* wreck. (Hill et al., Chap. 6, this volume)

High-resolution geospatial surveys are used to obtain an accurate representation of the wreck in three dimensions *in situ* on the seabed. A combination of subsea data acquisition techniques, including side-scan sonar, multibeam sonar and photogrammetry is used to obtain the data and integrated to create a high-resolution 3D model of the wreck (Fig. 11.2). Other measurements, such as hull thickness, can be used to determine the residual dimensions of the shell plating to help assess the potential for further structural degradation and hence the risk of pollutant release.

To produce the model, high-resolution survey data and hull thickness measurements are then integrated into a 3D digital environment to provide a better understanding of the wreck on the seabed and the likely locations of the pollutants within the wreck. This enables a 'virtual' structural inspection of the wreck, which can identify breaches in the hull and the presence of potential egress points for any pollutants. Figure 11.3 shows the integration of the digital model of the *Derbent* wreck highlighting the internal structure of the wreck and the locations of the cargo tanks. The integration of the data is critical to understanding the structural collapse that can be seen at the stern, noting that the structural failure is in line with the transverse bulkheads.

The integration of all these data provides a qualitative assessment, from which a range of possibilities of oil volumes remaining in the wreck can be estimated. To determine the oil volumes remaining within a wreck, some form of sampling regime is necessary, which means either intrusive or non-intrusive sampling must be undertaken. The 3D virtual model of the wreck also allows detailed planning of sample points for quantifying the pollutants remaining in the wreck.

Non-intrusive sampling uses specialised systems such as a Neutron Backscatter System (NBS), which can identify the level of hydrocarbons contained within a compartment, therefore enabling an assessment of the volume and location of hydrocarbons remaining in a wreck. The successful application of such systems, typically with an ROV, depends on the condition and orientation of the wreck, and it must be calibrated against known compartment conditions to provide results of sufficient veracity. Being non-intrusive, this method greatly reduces the risk of oil spills during the wreck assessment phase (therefore negating costly clean-up

Fig. 11.3 Image of the *Derbent* showing the digital model integrated with the multibeam sonar data

Fig. 11.4 NBS sampling regime and results overlaid on the integrated data model of the *Derbent*

resources and response) and has no impact on the structure of the wreck. Figure 11.4 shows the NBS readings on the wreck of SS *Derbent* (Hill et al., Chap. 6, this volume). The sampling regime has been designed to identify the water/oil interface within cargo tanks that have been identified as potentially containing hydrocarbons from the structural assessment.

Intrusive sampling requires tanks to be drilled and physically sampled. This can be helpful for any eventual oil removal operation, as valves can be installed that will provide samples and, if oil is identified, can then be used for the oil removal. Alternatively, for assessing the PPW, smaller scale sampling techniques can be used drilling small holes to 'tap off' relatively small volumes of oil from the tank, which are then sealed after sampling. Any intrusive sampling has a higher impact on the hull of the wreck, increasing the risk of an uncontrolled release of oil. The advantage of intrusive sampling is that the results provide positive, visual, confirmation of the existence of pollutants and samples collected can be sent for analysis to inform the intervention planning.

To help understand the application of the assessment methodology in practice, here we summarise two case studies for PPW assessments that we have carried out. Please also see the summaries of the assessments of the SS *Derbent* wreck and the RFA *War Mehtar* wreck by Hill and colleagues earlier in this volume (Chap. 6).

11.4 Case Study: Tug *Simson*

The tug *Simson* sank in the Finnish archipelago in Baltic Sea in 1978, after striking a submerged rock whilst towing a rock barge. The assessment of the *Simson* wreck was part of the PPW management programme being undertaken by the Finnish Environment Institute. The wreck was classified as potentially polluting as it contained an estimated 1000 litres of diesel fuel, and the wreck site is located within the Finnish Archipelago which is a conservation area protected under Finnish and European law. It was chosen as a test case to apply the proposed methodology, being smaller in size, accessible and with a lower intervention risk than other larger wrecks.

Waves Group undertook the wreck assessment to inform the future management plan and assess whether remediation of the wreck was necessary. On initial survey, it was found that the wreck had sunk into the mud in an upright position and was buried up to its gunwales. It was therefore not possible to collect high-resolution imagery of the hull and as such, the methodology had to be adapted, whilst on site.

The 3D wreck model, built using derived plans for the vessel and the initial survey data, became the data reference point. Instrumental in this was also interviews with the surviving Chief Engineer!

Detailed video footage of the wreck was fused into the 3D model to create an immersive model from which the above-seabed features of the wreck could be interrogated and assessed. Water and sediment samples were also taken to assess whether any diesel was leeching into the seabed or water column. Using this methodology, we were able to assess the condition of the wreck and assess whether the wreck was leaking hydrocarbon and the viability of further remediation measures. Figure 11.5 shows a view of the combined 3D model.

Based on these adapted data, the survey concluded there was likely to be only small amounts of the diesel oil remaining in the wreck. The methodology adopted demonstrates how the integration of data with the relevant salvage experience can be adapted and used rapidly to produce results with high confidence to inform the future management plan for the wreck.

Fig. 11.5 Tug *Simson* composite 3D model comprising multibeam sonar data and aligned solid modelling of the internal layout

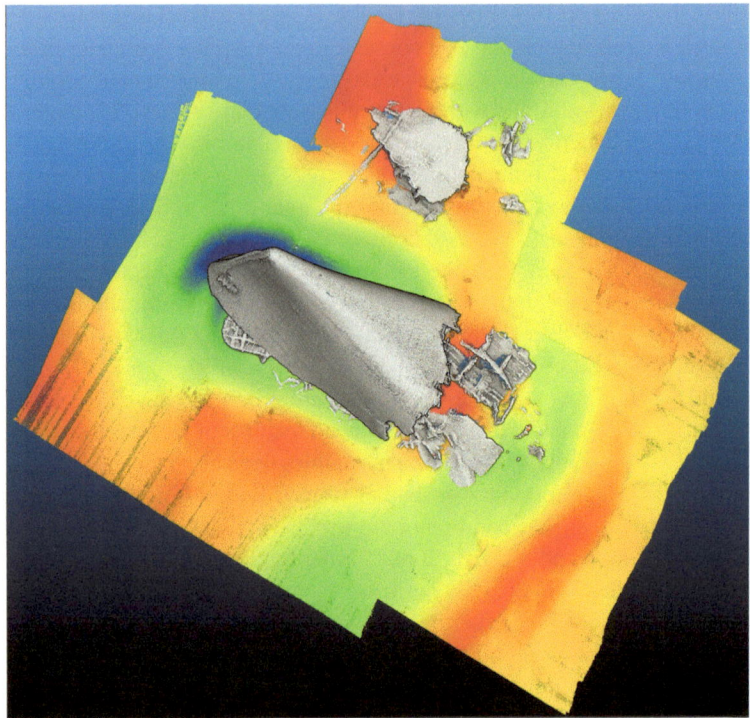

Fig. 11.6 Multibeam sonar image of unidentified vessel

11.4.1 Case Study: Emergency Response Assessment

It is, of course, preferred to conduct assessments of PPW in a proactive and planned manner. However, there are, and will be, occasions where a wreck starts to release pollutants before the planned assessment has been conducted.

Waves Group recently conducted a PPW assessment as an emergency response after significant oil was observed on the surface in a busy shipping area. Due to the rapid response required, the normal scoping phase could not be fully completed and mobilisation to site was conducted 'blind' with the consequence that the survey methodology had to be developed on site and evolved as findings were obtained.

The surface oil was traced to a wreck which was charted but unidentified (Fig.11.6). Temporary patching was installed to stop the release of oil and the wreck was surveyed using high-resolution multibeam sonar and combined with photo and video footage to create the 3D wreck model. An investigation was then undertaken to identify the vessel type, using archival data to compare and match key features to the survey data. Based on this, the vessel type was identified and the internal structural arrangement and compartments were incorporated into the 3D model (Fig 11.7). An analysis of the structural condition of the wreck was then

Fig. 11.7 Composite model comprising multibeam point cloud data a digital model of unidentified wreck showing internal compartments

conducted. Non-intrusive sampling of the oil was planned but could not be undertaken due to the poor condition of the wreck structure which could have resulted in the significant unintended release of oil.

An assessment of the risk of future release of hydrocarbons was then conducted, considering a range of factors. This assessment considered the potential volume of hydrocarbons in each tank (and their maximum capacity), the volumes reported in the pollution reports, the potential for material volumes to be recovered in the event of a further release and the condition of each compartment and whether it was already breached (i.e., was any oil already lost to sea). With the potential volume of oil remaining, an intervention strategy was recommended.

Emergency responses operations like this serve to highlight the need for a more proactive approach in the future, where PPW can be assessed, and the risks identified in a planned campaign. By using the similarities between vessel designs, assessments can be made with good confidence, even in rapid response cases without the time to plan the assessment. At the core of this, is the ability to adapt the assessment methodology.

11.5   Wreck Assessment Methodologies—The Future

The work that Waves Group has completed on PPW has brought a standardised approach to the assessment of PPW that has enabled stakeholders to obtain a clear understanding of the risks presented by a wreck, providing increased certainty on the risks it presents, and the intervention and management strategy needed. This methodology has proven to be readily adaptable between planned wreck assessments and rapid (or emergency) response cases. Figure 11.8 below sets out this methodology.

Phase one is typically a desk-based study undertaken to create a wreck inventory that identifies all the known PPW in the area of concern, followed by an appraisal of risk to prioritise further work on specific wrecks. Phase two is the assessment of the wrecks to establish whether the risk identified at the desk-based phase is appropriate. This phase will include the onsite survey work and is where our methodology has been effectively utilised. Phase three is the management and intervention phase which may include pollutant removal operations, containment, monitoring and the ongoing management of the cultural heritage, noting that many PPW are designated war graves.

As outlined here, there are a number of PPW risk assessment methodologies being utilised by different nations, with some regional approaches also in existence. However, the approaches between them differ, in terms of prioritisation of the wrecks, data acquisition and analysis for wreck assessments, and in intervention strategies. Moreover, there is no common approach between nations (and in particular, between those that have responsibility for the majority of PPW) on how to deal with the increasing pollution risk arising from PPW and who pays for it.

As the condition of PPW continues to deteriorate, there is an increasing need for a common approach to reduce the risk of serious marine pollution from PPW, with the potentially devastating effects that this will have on the ocean environment, as well as the socio-economic impacts.

Proactive management of PPW will require funding. Presently, even developed nations have limited budgets for managing PPWs and despite the increased awareness of the growing risk, are still reactive in dealing with those wrecks that are their responsibility. The wreck management programs that do exist are budget limited and so, are constrained in their ability to proactively manage the number of wrecks necessary.

Therefore, opening up pathways for alternative sources of funding will be critical to enabling the proactive management of PPW. For funding to become available, there will need to be certainty of the outcome; that is, what are the costs and what do the funders get for their money?

An international standardised methodology for the assessment and management of PPW, supported by appropriate international standards and guidelines, will provide that certainty and the confidence that the methodology being used represents best practice. That is, it enhances the due diligence for the funding. It allows the costs of PPW assessments to be more easily predicted and reduced (e.g., by sharing resources and undertaking assessments in campaigns, to a standard methodology). A standardised methodology will also support the aims of international treaties and regulations (e.g., the Nairobi Wreck Removal Convention 2007) and is directly linked to international treaties and targets such as:

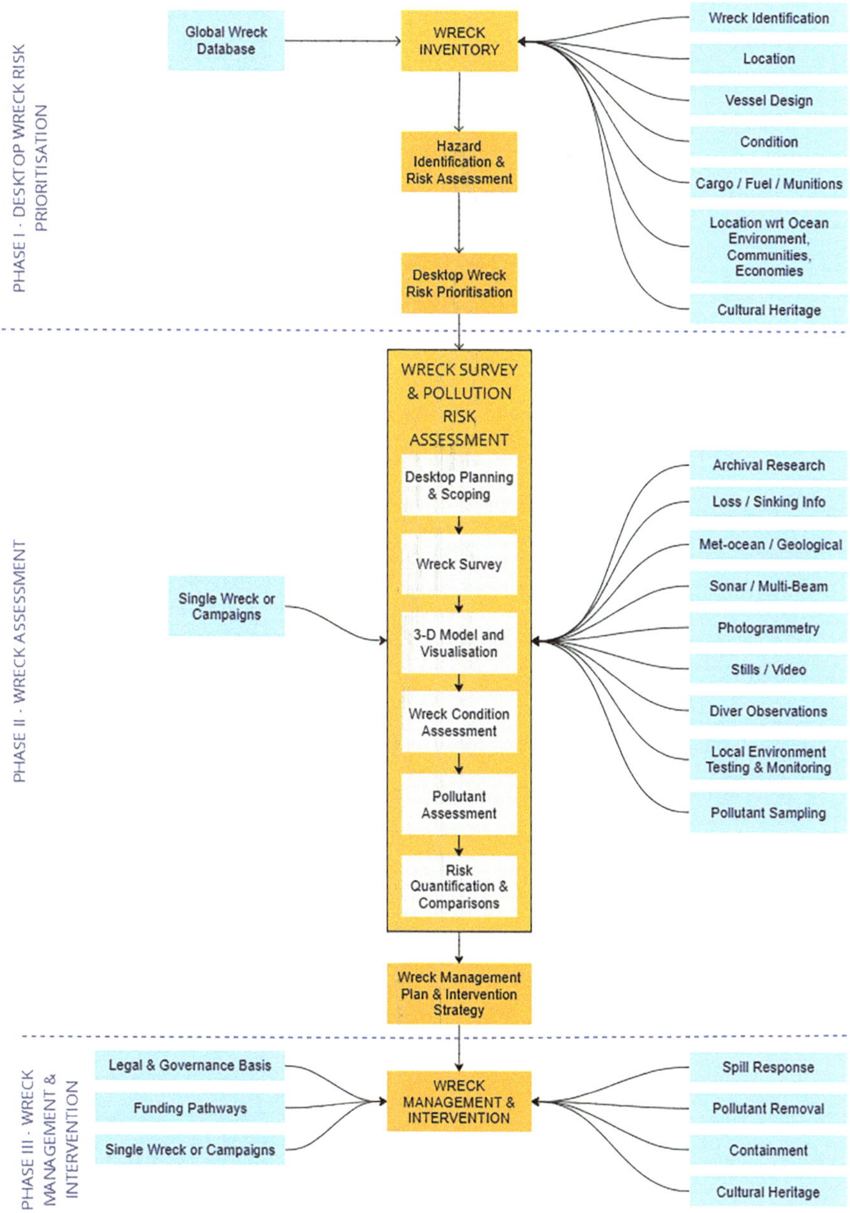

Fig. 11.8 Potentially polluting wreck assessment methodology

- The UN Sustainable Development Goal 14 targets to "prevent and significantly reduce marine pollution of all kinds by 2025".
- The UN Decade of Ocean Science (2021–2030), where an objective is "a clean ocean where sources of pollution are identified and reduced or removed".

- The UN High Seas Treaty (BBNJ), ratified in June 2023, which aims to protect the biodiversity of the ocean, through the creation of Marine Protected Areas (30% of the High Seas by 2030).

It is our goal to see the increased awareness of the PPW problem catalyse global action to develop an international standard and appropriate methodologies, drawing on existing international expertise, experience, and bringing technological innovation to help solve the problem. As we write this action has already commenced with the start of a project with Lloyd's Register Foundation, The Ocean Foundation and Waves Group called "*A Global Framework for the Near and Long Term Assessment, Intervention and Sharing of Data for Potentially Polluting Wrecks*".

References

Alcaro, L., Amato, E., Cabioch, F., Farchi, C., Gouriou, V. (2007). *DEEPP project: Development of European guidelines for potentially polluting shipwrecks*. ICRAM, Instituto Centrale per la Ricerca scientifica e tecnologica Applicata al Mare, CEDRE, Centre de Documentation de Recherché et d'Epérimentations sur les pollutions accidentelles des eaux.

Brennan, M. L., Delgado, J. P., Jozsef, A., Marx, D. E., & Bierwagen, M. (2023). Site formation processes and pollution risk mitigation of World War II oil tanker shipwrecks: *Coimbra* and *Munger T. Ball*. *Journal of Maritime Archaeology, 18*, 321–335.

Goodsir, F., Lonsdale, J. A., Mitchell, P. J., Suehring, R., Farcas, A., Whomersley, P., Brant, J. L., Clarke, C., Kirby, M. F., Skelhorn, M., & Hill, P. G. (2019). A standardised approach to the environmental risk assessment of potentially polluting wrecks. *Marine Pollution Bulletin, 142*, 290–302.

Hac, B., et al. (2019). Oil removal operations on baltic shipwrecks—Proposition of a wreck management programme for Poland. http://www.fundacjamare.pl/file/repository/2021MARE_WRECKS_GENERAL_METHODOLOGY_of_oil_removal_operations_REPORT_1.pdf

Hamer, M. (2010). Why wartime wrecks are slicking time bombs. *New Scientist*, Issue 2776.

Hill, P. G. (2022). Assessing the environmental risk posed by a legacy tanker wreck: A case study of the RFA War Mehtar. *Environmental Research Communications, 4*, 055005.

International Maritime Organization (IMO), Nairobi International Convention on the Removal of Wrecks, 2007, as amended.

Landquist, H., Hassellöv, I.-M., Rosén, L., Lindgren, J. F., & Dahllöf, I. (2013). Evaluating the needs of risk assessment methods of potentially polluting shipwrecks. *Journal of Environmental Management, 119*, 85–92.

Landquist, H., et al. (2016). VRAKA—A probabilistic risk assessment method for potentially polluting shipwrecks. *Frontiers in Environmental Science, 4*, 49.

Monfils, R. (2005). The global risk of marine pollution from WWII shipwrecks: Examples from the Seven Seas. In *International Oil Spill Conference, IOSC 2005*. Miami Beach.

NOAA. (2013). Risk assessment for potentially polluting wrecks in U.S. *Waters*. 195 pp, https://nmssanctuaries.blob.core.windows.net/sanctuaries-prod/media/archive/protect/ppw/pdfs/2013_potentiallypollutingwrecks.pdf

SPREP. (2002). A Regional Strategy to address Marine Pollution from World War II Wrecks. 13th SPREP meeting, Majuro, Marshall Islands, July 2002.

SYKE. (2016). *Sunken Wreck Environmental Risk Assessment (SWERA)*. https://www.syke.fi/projects/swera

Open Access This chapter is licensed under the terms of the Creative Commons Attribution 4.0 International License (http://creativecommons.org/licenses/by/4.0/), which permits use, sharing, adaptation, distribution and reproduction in any medium or format, as long as you give appropriate credit to the original author(s) and the source, provide a link to the Creative Commons license and indicate if changes were made.

The images or other third party material in this chapter are included in the chapter's Creative Commons license, unless indicated otherwise in a credit line to the material. If material is not included in the chapter's Creative Commons license and your intended use is not permitted by statutory regulation or exceeds the permitted use, you will need to obtain permission directly from the copyright holder.

Chapter 12
Concluding Statement

Joao Sousa, Benjamin Ferrari, Michael L. Brennan, and Ole Varmer

The grim reality of potentially polluting legacy wrecks presents a multifaceted threat that extends far beyond the decaying hulks on the ocean floor. These wrecks serve as silent sentinels of a bygone era, leaking toxic substances that indiscriminately poison marine life and disrupt the intricate web of oceanic relationships. From the smallest plankton to the largest apex predators, no organism remains untouched by the effects of these pollution sources.

The toxicity seeping from these underwater time bombs endangers the delicate balance of biodiversity in our oceans. It also poses a profound threat to the safety of people and ocean infrastructure. It jeopardises the resilience and livelihoods of communities that depend on these marine ecosystems. Coastal communities, fishing industries, and tourism sectors all bear the brunt of this environmental degradation, facing economic and social consequences that will resonate for generations. What is more, polluting legacy wrecks pose a material risk to planned strategic investment in protection and restoration of ecosystems through financial mechanisms such as Blue Bonds (a subset of the green, social and sustainable bonds), and management tools such as MPAs.

J. Sousa
International Union for Conservation of Nature, Gland, Switzerland

B. Ferrari
Lloyd's Register Foundation, Bath, UK

M. L. Brennan (✉)
SEARCH Inc., Jacksonville, FL, USA
e-mail: mike@brennanexploration.com

O. Varmer
The Ocean Foundation, Washington, DC, USA

© The Author(s) 2024   157
M. L. Brennan (ed.), *Threats to Our Ocean Heritage: Potentially Polluting Wrecks*, SpringerBriefs in Underwater Archaeology,
https://doi.org/10.1007/978-3-031-57960-8_12

There are substantial barriers that must be overcome in order to move away from the current reliance on emergency response interventions to a more strategic approach that delivers assurance of long-term safety and efficient use of resources. In terms of governance there is a lack of clarity regarding responsibility and liability. While many legacy wrecks have been located, many have yet to be found and particular challenges will be posed by those in deep water where remediation efforts become even more challenging. Regardless of location, historic preservation law and policy must be factored in. That may also call for sensitivity of grave sites and mitigation to avoid or minimise adverse impacts to cultural heritage. The heritage should be identified, considered in EIAs, and addressed in planning and management. A survey should be completed and mitigation considered before intrusive research and/or recovery. For PPWs that are also grave sites, a memorial service or a plaque may be appropriate. In addressing the duty to protect our heritage and the marine environment, it is understood that priority be given to a clean ocean particularly for those livelihoods so dependent upon it. If it is not safe for life, property, and the marine environment, it is not sustainable. A lack of widely accepted technical standards and protocols inhibits engineering intervention and strategic planning. In the absence of such standards, it is also harder to advocate for action to aggregate the necessary financial and multi-lateral agency support for change.

It is evident that a proactive and comprehensive approach is needed. It is essential for governments, industries, and environmental organisations to collaborate on strategies for identifying, assessing, and mitigating the risks associated with polluting wrecks. Whether through the removal of hazardous materials, the implementation of protective measures, the development of sustainable restoration initiatives or creating a relief fund to support lost livelihoods, there is a collective responsibility to safeguard our oceans and the life they support.

Public awareness and engagement are also vital components of any successful solution. Communities that rely on marine resources must be informed and empowered to advocate for the preservation of their environment. Education campaigns, research efforts, international cooperation and policy development will be instrumental in addressing the multifaceted challenges posed by polluting wrecks.

The urgency of addressing this impending crisis cannot be overstated. New resources are required—especially in the Global South—to boost resilience and manage risk. We believe that these resources, technical and financial, will only become available at the level required once there is a clear and widely accepted roadmap to better governance and optimised interventions. To this end, at the time of publication, Lloyd's Register Foundation is supporting partner organisations to establish an evidence base, including this volume, that will underpin advocacy for strategic action. The Ocean Foundation and Waves Group will coordinate a series of international expert workshops to forge consensus on the change that we need. The IUCN and ICOMOS are also involved and will be instrumental in enhancing cooperation with the UN family of agencies in order to marshal support for new international regulations and protocols.

In the face of this ecological and socio-economic challenge, our responsibility is clear. By prioritising the protection of our oceans and the preservation of the myriad species that call it home, we can not only safeguard biodiversity but also ensure the safety and sustainability of livelihoods that depend on these precious marine resources. The time to act is now, for the sake of our planet, its inhabitants, and the generations to come.

Open Access This chapter is licensed under the terms of the Creative Commons Attribution 4.0 International License (http://creativecommons.org/licenses/by/4.0/), which permits use, sharing, adaptation, distribution and reproduction in any medium or format, as long as you give appropriate credit to the original author(s) and the source, provide a link to the Creative Commons license and indicate if changes were made.

The images or other third party material in this chapter are included in the chapter's Creative Commons license, unless indicated otherwise in a credit line to the material. If material is not included in the chapter's Creative Commons license and your intended use is not permitted by statutory regulation or exceeds the permitted use, you will need to obtain permission directly from the copyright holder.